MECHANICAL ENGINEERING THEORY AND APPLICATIONS

ADDITIVE MANUFACTURING

COSTS, COST EFFECTIVENESS AND INDUSTRY ECONOMICS

MECHANICAL ENGINEERING THEORY AND APPLICATIONS

Additional books in this series can be found on Nova's website under the Series tab.

Additional e-books in this series can be found on Nova's website under the e-book tab.

MECHANICAL ENGINEERING THEORY AND APPLICATIONS

ADDITIVE MANUFACTURING

COSTS, COST EFFECTIVENESS AND INDUSTRY ECONOMICS

FELIPE BREWER
EDITOR

nova publishers
New York

Library of Congress Cataloging-in-Publication Data

ISBN: 978-1-63483-364-6

Published by Nova Science Publishers, Inc. † New York

CONTENTS

PREFACE

The use of additive manufacturing has increased significantly in previous years. Additive manufacturing is used by multiple industry subsectors, including motor vehicles, aerospace, machinery, electronics, and medical products. Currently, however, additive manufactured products represent less than one percent of all manufactured products in the U.S. As the costs of additive manufacturing systems decrease, this technology may change the way that consumers interact with producers. Additive manufacturing technology opens up new opportunities for the economy and society. It can facilitate the customized production of strong light-weight products and it allows designs that were not possible with previous manufacturing techniques. This book provides aggregate manufacturing industry data and industry subsector data to develop a quantitative depiction of the U.S. additive manufacturing industry.

In: Additive Manufacturing
Editor: Felipe Brewer

ISBN: 978-1-63483-364-6

Chapter 1

COSTS AND COST EFFECTIVENESS OF ADDITIVE MANUFACTURING: A LITERATURE REVIEW AND DISCUSSION*

Douglas S. Thomas and Stanley W. Gilbert

ABSTRACT

The use of additive manufacturing has increased significantly in previous years. Additive manufacturing is used by multiple industry subsectors, including motor vehicles, aerospace, machinery, electronics, and medical products. Currently, however, additive manufactured products represent less than one percent of all manufactured products in the U.S. As the costs of additive manufacturing systems decrease, this technology may change the way that consumers interact with producers. Additive manufacturing technology opens up new opportunities for the economy and society. It can facilitate the customized production of strong light-weight products and it allows designs that were not possible with previous manufacturing techniques. Various challenges, however, can impede and slow the adoption of this technology. In many instances, the cost of roducing a product using additive manufacturing processes exceeds that of traditional methods. This report examines literature on the costs of additive manufacturing and seeks to identify those instances where additive manufacturing might be cost effective and also identify potential means for reducing costs when using this technology. Current

* This is an edited, reformatted and augmented version of NIST Special Publication 1176, issued by the National Institute of Standards and Technology, December 2014.

research on additive manufacturing costs reveals that this technology is cost effective for manufacturing small batches with continued centralized manufacturing; however, with increased automation distributed production may become cost effective. Due to the complexities of measuring additive manufacturing costs, current studies are limited in their scope. Many of the current studies examine the production of single parts. Those that examine assemblies tend not to examine supply chain effects such as inventory and transportation costs along with decreased risk to supply disruption. Currently, research also reveals that material costs constitute a major proportion of the cost of a product produced using additive manufacturing. However, technologies can often be complementary, where two technologies are adopted alongside each other and the benefits are greater than if they were adopted individually. Increasing adoption of additive manufacturing may lead to a reduction in raw material cost through economies of scale. The reduced cost in raw material might then propagate further adoption of additive manufacturing. There may also be economies of scale in raw material costs if particular materials become more common rather than a plethora of different materials.

The additive manufacturing system is also a significant cost factor; however, this cost has continually decreased. Between 2001 and 2011 the average price decreased 51% after adjusting for inflation.

PREFACE

This study was conducted by the Applied Economics Office in the Engineering Laboratory at the National Institute of Standards and Technology. The study provides aggregate manufacturing industry data and industry subsector data to develop a quantitative depiction of the U.S. additive manufacturing industry.

1. INTRODUCTION

1.1. Background

In 2011, the world produced approximately $11.3 trillion in manufacturing value added, according to United Nations Statistics Division (UNSD) data. The U.S. produced approximately 17% of these goods, making it the second largest manufacturing nation in the world, down from being the largest in 2009. Many products and parts made by the industry are produced by taking pieces of raw material and cutting away sections to create the desired part or by injecting material into a mold; however, a relatively new process called additive manufacturing is beginning to take hold where material is aggregated together rather than formed in a mold or cut away. Additive manufacturing is the process of joining materials to make objects from three-dimensional (3D) models layer by layer as opposed to subtractive methods that remove material.

The terms additive manufacturing and 3D printing tend to be used interchangeably to describe the same approach to fabricating parts. This technology is used to produce models, prototypes, patterns, components, and parts using a variety of materials including plastic, metal, ceramics, glass, and composites. Products with moving parts can be printed such that the pieces are already assembled. Technological advances have even resulted in a 3D-Bio-printer that one day might create body parts on demand.[1,2]

Additive manufacturing is used by multiple industry subsectors, including motor vehicles, aerospace, machinery, electronics, and medical products.[3] This technology dates back to the 1980's with the development of stereolithography, which is a process that solidifies layers of liquid polymer using a laser. The first additive manufacturing system available was the SLA-1 by 3D Systems. Technologies that enabled the advancement of additive manufacturing were the desktop computer and the availability of industrial lasers.

Although additive manufacturing allows the manufacture of customized and increasingly complex parts, the slow print speed of additive manufacturing systems limits their use for mass production. Additionally, 3D scanning technologies have enabled the replication of real objects without using expensive molds.

As the costs of additive manufacturing systems decrease, this technology may change the way that consumers interact with producers. The customization of products will require increased data collection from the end user. Additionally, an inexpensive 3D printer allows the end user to produce polymer-based products in their own home or office. Currently, there are a number of polymer systems that are within the budget of the average consumer.

Globally, an estimated $967 million in revenue was collected for additive manufactured goods[4] with the U.S. accounting for an estimated $367 million or 38% of global production in 2013.[5]

Table 1.1 provides a comparison of additive manufactured products and total industry production for 2011. Additive manufactured products are categorized as being in the following sectors: motor vehicles; aerospace; industrial/business machines; medical/dental; government/military; architectural; and consumer products/electronics, academic institutions, and other. The consensus among well- respected industry experts is that the penetration of the additive manufacturing market was 8% in 2011;[6] however, as seen in Table 1.1, goods produced using additive manufacturing methods represent between 0.01% and 0.05% of their relevant industry subsectors. Thus, additive manufacturing has sufficient room to grow.

There have been three proposed alternatives for the diffusion of additive manufacturing. The first is considered by many to be the most extreme where a significant proportion of consumers purchase additive manufacturing systems or 3D printers and produce products themselves.[7] The second is a copy shop scenario, where individuals submit their designs to a service provider that produces it.[8] Both of these scenarios are considered by many to be somewhat less likely.[9] The third scenario involves additive manufacturing being adopted by the commercial manufacturing industry, changing the technology of design and production. Additive manufacturing is seen as a practical alternative for commercial manufacturing in high wage economies, making it an opportunity for advancing U.S. manufacturing while maintaining and advancing U.S. innovation.

The U.S. is currently a major user of additive manufacturing technology and the primary producer of additive manufacturing systems. Approximately 62.8% of all commercial/industrial units sold in 2011 were made by the top three producers of additive manufacturing systems: Stratasys, Z Corporation, and 3D Systems based out of the United States.[10] Approximately 64.4% of all systems were made by companies based in the United States. If additive manufacturing has a saturation level between 5% and 35% of the relevant sectors, it is forecasted that it might reach 50% of market potential between 2031 and 2038, while reaching near 100% between 2058 and 2065. The industry would reach $50 billion between 2029 and 2031, while reaching $100 billion between 2031 and 2044.[11]

Table 1.1. Additive Manufacturing Shipments, 2011

Category	Relevant NAICS Codes	Percent of Total AM Made Products	Shipments of US Made AM Products ($millions, 2011)*	Total Shipments ($millions, 2011)	AM Share of Industry Shipments
Motor vehicles	NAICS 3361, 3362, 3363	19.5%	48.0	445 289.4	0.01%
Aerospace	NAICS 336411, 336412, 336413	12.1%	29.8	157 700.7	0.02%
Industrial/business machines	NAICS 333	10.8%	26.6	365 734.8	0.01%
Medical/dental	NAICS 3391	15.1%	37.2	89 519.5	0.04%
Government/military	NAICS 336414, 336415, 336419, 336992	6.0%	14.8	32 784.4	0.05%
Architectural	NAICS 3323	3.0%	7.4	72 186.9	0.01%
Consumer products/electronics, academic institutions, and other	All other within NAICS 332 through 339	33.6%	82.7	895 709.8	0.01%
TOTAL	NAICS 332 through 339	100.0%	246.1	2 058 925.5	0.01%

* These values are calculated assuming that the percent of total additive manufacturing made products for each industry is the same for the U.S. as it is globally. It is also assumed that the U.S. share of AM systems sold is equal to the share of revenue for AM products.

Note: Numbers may not add up to total due to rounding.

1.2. Purpose

Additive manufacturing technology opens up new opportunities for the economy and society. It can facilitate the production of strong light-weight products for the aerospace industry and it allows designs that were not possible with previous manufacturing techniques. It may revolutionize medicine with biomanufacturing. This technology has the potential to increase the well-being of U.S. citizens and improve energy efficiency in ground and air transportation.

However, the adoption and diffusion of this new technology is not instantaneous. With any new technology, new standards, knowledge, and infrastructure are required to facilitate its use. Organizations such as the National Institute of Standards and Technology can enable the development of these items; thus, it is important to understand the costs and benefits of the additive manufacturing industry. This report examines literature on the costs of additive manufacturing and seeks to identify areas where it maintains a cost advantage and identify potential areas for cost reductions.

1.3. Scope and Approach

This report focuses on the costs of additive manufacturing; however, many of the advantages of additive manufacturing may lie in improvements of the finished good. Therefore, there is some discussion on the product improvements that result from additive manufacturing technologies. Section 2 provides an overview of the processes and materials used in additive manufacturing. It also discusses the literature on additive manufacturing costs and categorizes them by their process and material combination. Section 3 provides a discussion and examination of the costs and benefits of additive manufacturing. It is broken into ill-structured costs, well-structured costs, and product enhancements and quality.

Section 4 provides an examination of the cost models used to examine additive manufacturing. Section 5 provides a discussion on the trends in implementation and adoption of additive manufacturing.

2. Additive Manufacturing Processes, Materials, and Literature

There are a number of additive manufacturing processes; however, at first glance it may appear that there are more types than in actuality. Many companies have created unique system and material names in order to differentiate themselves, which has created some confusion. Fortunately, there has been some effort to categorize the processes and materials using standard methods. The categorization and descriptions of processes and materials below relies heavily on Wohlers (2012) and ASTM International Standards.[12]

2.1. Processes

The total global revenue from additive manufacturing system sales was $502.5 million with U.S. revenue estimated at $323.6 million. These systems are categorized into various different processes. ASTM International Committee F42.91 on Additive Manufacturing Technologies has developed standard terminologies. Provided below are the categories and adapted definitions from the ASTM F2792 standard:

Binder Jetting: This process uses liquid bonding agent deposited using an inkjet-print head to join powder materials in a powder bed.

Directed Energy Deposition: This process utilizes thermal energy, typically from a laser, to fuse materials by melting them as they are deposited.

Material Extrusion: These machines push material, typically a thermoplastic filament, through a nozzle onto a platform that moves in horizontal and vertical directions.

Material Jetting: This process, typically, utilizes a moving inkjet-print head to deposit material across a build area.

Powder Bed Fusion: This process uses thermal energy from a laser or electron beam to selectively fuse powder in a powder bed.

Sheet Lamination: This process uses sheets of material bonded to form a three- dimensional object.

Vat Photopolymerization: These machines selectively cure a liquid photopolymer in a vat using light.

2.2. Materials

Approximately $327.1 million was spent globally on materials for additive manufacturing in 2011.[13] There are two primary types of materials: plastics and metals. There are also ceramics, composites, and other materials that are used as well, but are not as common. Wohlers groups the materials into eight categories:

- Polymers and polymer blends
- Composites
- Metals
- Graded/hybrid metals
- Ceramics
- Investment casting patterns
- Sand molds and cores
- Paper

Certain processes lend themselves to certain materials. Table 2.1 presents the combinations of additive manufacturing processes and their corresponding materials. The combinations that are left blank are material/process combinations that are not currently utilized.

2.3. Cost Literature

There are two major motivational categories for examining additive manufacturing costs. The first is to compare additive manufacturing processes to other traditional processes such as injection molding and machining. The purpose of these types of examinations is to determine under what circumstances additive manufacturing is cost effective. The second category involves identifying resource use at various steps in the additive manufacturing process. The purpose of this type of analysis is to identify when and where resources are being consumed and whether there can be a reduction in resource use. Table 2.2 provides a literature list for cost studies on additive manufacturing categorized by the combinations of additive manufacturing processes and corresponding materials shown in Table 2.1. The areas in black are those areas that are not possible (i.e., they are the empty cells from Table 2.1) while those with an "x" indicate possible combinations but no cost literature was identified. One column has been added to indicate studies that examine both additive manufacturing and traditional manufacturing. The documents listed in the table are heavily relied on for characterizing the costs of additive manufacturing. Two major components that affect costs are the build time and the energy consumption of additive manufacturing systems. Although these issues will not be discussed at significant length, a selection of literature is categorized in Table 2.3 and Table 2.4.

3. ADDITIVE MANUFACTURING COSTS AND BENEFITS

As discussed by Young (1991), the costs of production can be categorized in two ways.[14] The first involves those costs that are "well-structured" such as labor, material, and machine costs. The second involves "ill-structured costs" such as those associated with build failure, machine setup, and inventory. In the literature, there tends to be more focus on well-structured costs of additive manufacturing than ill-structured costs; however, some of the more significant benefits and cost savings in additive manufacturing may be hidden in the ill-structured costs. Moreover considering additive manufacturing in the context of lean production might be useful.

A key concept of lean manufacturing is the identification of waste, which is classified into seven categories:

1) Overproduction: occurs when more is produced than is currently required by customers
2) Transportation: transportation does not make any change to the product and is a source of risk to the product
3) Rework/Defects: discarded defects result in wasted resources or extra costs correcting the defect
4) Over-processing: occurs when more work is done than is necessary
5) Motion: unnecessary motion results in unnecessary expenditure of time and resources
6) Inventory: is similar to that of overproduction and results in the need for additional handling, space, people, and paperwork to manage extra product
7) Waiting: when workers and equipment are waiting for material and parts, these resources are being wasted

Table 2.1. Additive Manufacturing Process and Material Combinations

	Material extrusion	Material jetting	Binder jetting	Vat photopoly-merization	Sheet lamination	Powder bed fusion	Directed energy deposition
Polymers and polymer blends	x	x	x	x	x	x	
Composites		x	x	x		x	
Metals		x	x		x	x	x
Graded/hybrid metals					x		x
Ceramics			x	x		x	
Investment casting patterns		x	x	x		x	
Sand molds and cores	x		x			x	
Paper					x		

Source: Wohlers, Terry. "Wohlers Report 2012: Additive Manufacturing and 3D Printing State of the Industry." Wohlers Associates, Inc. 2012.

Table 2.2. Literature on the Costs of Additive Manufacturing

	Material extrusion	Material jetting	Binder jetting	Vat photopoly-merization	Sheet lamination	Powder bed fusion	Directed energy deposition	Additive Manufacturing research that includes Traditional Manufacturing
Polymers, polymer blends, and composites	T.A. Grimm (2010)*; Hopkinson and Dickens (2003); Hopkinson (2006); Baumers (2012)	T.A. Grimm (2010)*	T.A. Grimm (2010)*	T.A. Grimm (2010)*; Hopkinson and Dickens (2003); Hopkinson (2006); Li (2005)	T.A. Grimm (2010)*	Ruffo, Tuck, and Hague (2006a); Baldinger and Duchi (2013); Ruffo and Hague (2007); Hopkinson and Dickens (2003); Hopkinson (2006); Baumers (2012); Zhang and Bernard (2014); Atzeni et al. (2010)		Hopkinson (2006); Ruffo, Tuck, and Hague (2006a); Ruffo and Hague (2007); Hopkinson and Dickens (2003); Atzeni et al. (2010); Li (2005)
Metals		x	x		x	Rickenbacher (2013)*; Baumers et al. (2012); Baumers (2012); Baumers et al (2013); Atzeni, Iuliano and Salmi (2011); Atzeni and Salmi (2012); Lindemann et al. (2012); Lindemann et al. (2013)	x	Allen (2006)
Graded/hybrid metals					x		x	
Ceramics			x	x		x		
Investment casting patterns		x	x	x		x		
Sand molds and cores	x		x			x		
Paper					x			
Undesignated Material						Khajavi et al. (2014)		

* 3D Printing.

Table 2.3. Literature on the Build Time of Additive Manufacturing

	Material extrusion	Material jetting	Binder jetting	Vat photopoly-merization	Sheet lamination	Powder bed fusion	Directed energy deposition	Undesignated Process	Additive Manufacturing research that includes Traditional Manufacturing
Polymer, polymer blend, and composites	x	x	x	Giannatsis et al. (2001); Chen and Sullivan (1996); Campbell (2008)	x	Choi and Samavedam (2002); Ruffo et al. (2006b); Munguia et al. (2009); Zhang and Bernard (2014); Kruth et al. (2005)			
Metals		x	x		x	Rickenbacher et al. (2013)	x		
Graded/hybrid metals					x		x		
Ceramics			x	x		x			
Investment casting patterns		x	x	x		x			
Sand molds and cores	x		x			x			
Paper					x				
Undesignated Material								Di Angelo and Di Stefano (2011)	

Table 2.4. Literature on the Energy Consumption of Additive Manufacturing

	Material extrusion	Material jetting	Binder jetting	Vat photopoly-merization	Sheet lamination	Powder bed fusion	Directed energy deposition	Undesignated Process	Additive Manufacturing research that includes Traditional Manufacturing
Polymers, polymer blends, and composites	Mognol et al. (2006); Baumers (2011); Luo et al. (1999)	x	x	Mognol et al. (2006); Luo et al. (1999)	x	Sreenivasan et al. (2009); ATKINS (2007); Telenko and Seepersad (2010); Telenko and Seepersad (2012); Paul and Anand (2012); Kellens et al. (2011);Verma and Rai; Baumers (2011); Sreenivasan and Bourell (2009); Baumers et al. (2010); ; Luo et al. (1999); Kellens et al. (2010); Sreenivasan et al. (2010)			ATKINS (2007); Telenko and Seepersad (2012)
Metals		x	x		x	Baumers et al. (2012); Baumers et al (2013); Mognol et al. (2006); Baumers (2011); Sreenivasan et al. (2010)	Morrow et al. (2007)	ATKINS Project (2007)	ATKINS Project (2007); Morrow et al. (2007)
Graded/hybrid metals					x		x		
Ceramics			x	x		x			
Investment casting patterns		x	x	x		x			
Sand molds and cores	x		x			x			
Paper					x				

Additive manufacturing may impact a significant number of these categories. For example, additive manufacturing may significantly reduce the need for large inventory, which is a significant cost in manufacturing. In 2011, there was an average of $208 billion or the equivalent of 14% of annual revenue held in inventory for medium- and high-tech manufacturing[15] with an estimated cost of $52 billion or 3% of revenue.[16] Reducing inventory frees up capital and reduces expenses. The following sections will attempt to discuss some of the potential savings and benefits of additive manufacturing as well as its costs.

3.1. Ill-Structured Costs

Many costs are hidden in the supply chain, which is a system that moves products from supplier to customer. Additive manufacturing may, potentially, have significant impacts on the design and size of this system, reducing its associated costs.[17]

3.1.1. Inventory and Transportation

Inventory: At the beginning of 2011, there were $537 billion in inventories in the manufacturing industry, which was equal to 10% of that year's revenue. The resources spent producing and storing these products could have been used elsewhere if the need for inventory were reduced. Suppliers often suffer from high inventory and distribution costs. Additive manufacturing provides the ability to manufacture parts on demand. For example, in the spare parts industry, a specific type of part is infrequently ordered; however, when one is ordered, it is needed quite rapidly, as idle machinery and equipment waiting for parts is quite costly. Traditional production technologies make it too costly and require too much time to produce parts on demand. The result is a significant amount of inventory of infrequently ordered parts.[18] This inventory is tied up capital for products that are unused. They occupy physical space, buildings, and land while requiring rent, utility costs, insurance, and taxes. Meanwhile the products are deteriorating and becoming obsolete. Being able to produce these parts on demand using additive manufacturing reduces the need for maintaining large inventory and eliminates the associated costs.

Transportation: Additive manufacturing allows for the production of multiple parts simultaneously in the same build, making it possible to produce an entire product. Traditional manufacturing often includes production of parts at multiple locations, where an inventory of each part might be stored. The parts are shipped to a facility where they are assembled into a product, as illustrated in Figure 3.1. Additive manufacturing has the potential to replace some of these steps for some products, as this process might allow for the production of the entire assembly. This would reduce the need to maintain large inventories for each part of one product. It also reduces the transportation of parts produced at varying locations and reduces the need for just-in-time delivery.

3.1.2. Consumer's Proximity to Production

As previously discussed, three alternatives have been proposed for the diffusion of additive manufacturing. The first is where a significant proportion of consumers purchase additive manufacturing systems or 3D printers and produce products themselves.[19] The second is a copy shop scenario, where individuals submit their designs to a service provider that produces goods.[20] The third scenario involves additive manufacturing being adopted by the commercial manufacturing industry, changing the technology of design and production. One might consider a fourth scenario. Because additive manufacturing can produce a final product in one build, there is limited exposure to hazardous conditions, and there is little hazardous waste,[21] there is the potential to bring production closer to the consumer for some products (i.e., distributed manufacture). For example, currently, a more remote geographic area may order automotive parts on demand, which may take multiple days to be delivered. Additive manufacturing might allow some of these parts or products to be produced near the point of use or even onsite.[22] Further, localized production combined with simplified processes may begin to blur the line between manufacturers, wholesalers, and retailers as each could potentially produce products in their facilities.

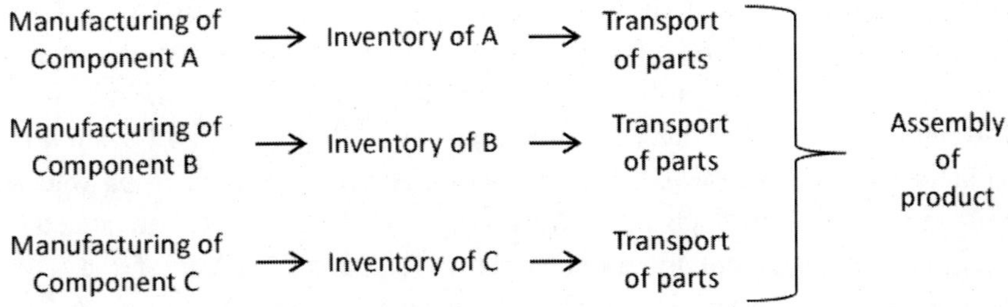

Figure 3.1. Example of Traditional Manufacturing Flow.

Khajavi et al. (2014) compare the operating cost of centralized additive manufacturing production and distributed production, where production is in close proximity to the consumer.[23] This analysis examined the production of spare parts for the air-cooling ducts of the environmental control system for the F-18 Super Hornet fighter jet, which is a well-documented instance where additive manufacturing has already been implemented. The expected total cost per year for centralized production was $1.0 million and $1.8 million for distributed production. Inventory obsolescence cost, initial inventory production costs, inventory carrying costs, and spare parts transportation costs are all reduced for distributed production; however, significant increases in personnel costs and the initial investment in additive manufacturing machines make it more expensive than centralized production. Increased automation and reduced machine costs are needed for this scenario to be cost effective. It is also important to note that this analysis examined the manufacture of a relatively simple component with little assembly. One potential benefit of additive manufacturing might be to produce an assembled product rather than individual components. Research by Holmström et al. (2010), which also examines spare parts in the aircraft industry, concurs that, currently, on demand centralized production of spare parts is the most likely approach to succeed; however, if additive manufacturing develops into a widely adopted process, the distributed approach becomes more feasible.[24]

3.1.3. Supply Chain Management

The supply chain includes purchasing, operations, distribution, and integration. Purchasing involves sourcing product suppliers. Operations involve demand planning, forecasting, and inventory. Distribution involves the movement of products and integration involves creating an efficient supply chain.25 Reducing the need for these activities can result in a reduction in costs. Some large businesses and retailers largely owe their success to the effective management of their supply chain. They have used technology to innovate the way they track inventory and restock shelves resulting in reduced costs. Walmart, for example, cut links in the supply chain, making the link between their stores and the manufacturers more direct. It also began vender managed inventory (VMI), where manufacturers were responsible for managing their products in Walmart's warehouses. It advanced its communication and collaboration network. The management of the supply chain can be the factor that drives a company to market leadership. Additive manufacturing may have significant impacts on the manufacturing supply chain, reducing the need for supply chain management. This technology has the potential to bring manufacturers closer to consumers, reducing the links in the supply chain.

3.1.4. Vulnerability to Supply Disruption

If additive manufacturing reduces the number of links in the supply chain and brings production closer to consumers, it will result in a reduction in the vulnerability to disasters and disruptions. Every factory and warehouse in the supply chain for a product is a potential point where a disaster or disruption can stop or hinder the production and delivery of a product. A smaller supply chain with fewer links means there are fewer points for potential disruption. Additionally, if production is brought closer to consumers it will result in more decentralized production where many facilities are producing a few products rather than a few facilities producing many products. Disruptions in the supply chain might result in localized impacts rather than regional or national impacts.

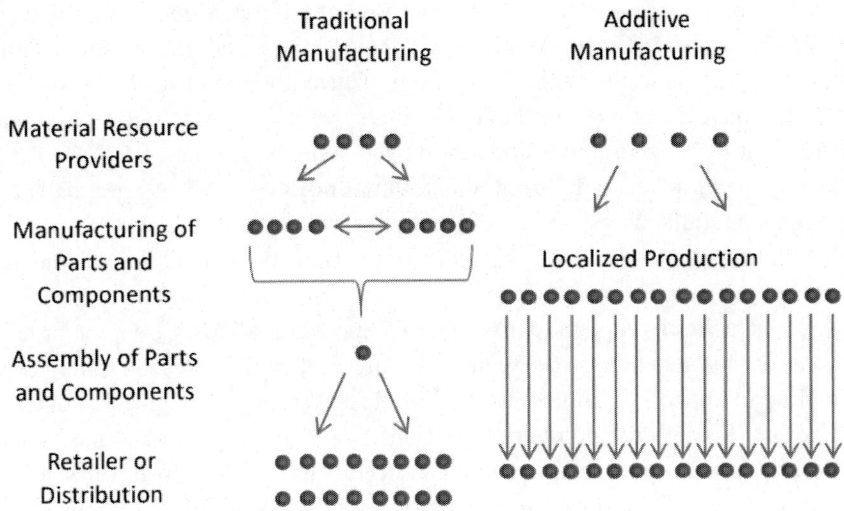

Figure 3.2. Example of Traditional Supply Chain Compared to the Supply Chain for Additive Manufacturing with Localized Production.

Figure 3.2 provides an example that compares traditional manufacturing to additive manufacturing. Under traditional manufacturing, material resource providers deliver to the manufacturers of parts and components, who might deliver parts and components to each other and then to an assembly plant. From there the assembled product is delivered to a retailer or distributer. A disruption at any of the points in manufacturing or assembly may result in a disruption of deliveries to all the retailers or distributers if there is not redundancy in the system. Additive manufacturing with localized production does not have the same vulnerability. First, there may not be any assembly of parts or components. Second, a disruption to manufacturing does not impact all of the retailers and distributers.

3.2. Well-Structured Costs

3.2.1. Material Costs

With geometric freedom, additive manufacturing allows products to be produced using less material while maintaining the necessary performance. Products can be produced at the level of performance needed rather than significantly exceeding the necessary performance level because of limitations in traditional manufacturing. Currently, however, the price of materials for additive manufacturing can often exceed those of traditional manufacturing.

Metal Material Costs: As discussed previously, metal and plastic are the primary materials used for this technology. Currently, the cost of material for additive manufacturing can be quite high when compared to traditional manufacturing. Atzeni and Salmi (2011) showed that the material costs for a selected metal part made from aluminum alloys was €2.59 per part for traditional manufacturing and €25.81 per part for additive manufacturing using selective laser sintering; thus, the additive manufacturing material was nearly ten times more expensive.[26]

Other research on metal parts confirms that material costs are a major cost driver for this technology as seen in Figure 3.3, which presents data for a sample part made of stainless steel. For this example, four cost factors are varied and the production quantity is a little less than 200 for the base case. This analysis provides insight into identifying the largest costs of additive manufacturing. The first cost factor that is varied is the building rate, which is the speed at which the additive manufacturing system operates. In this example, it is measured in cubic centimeters per hour. The second factor that is varied is the machine utilization measured as the number of hours per year that the machine is operated. The third factor is the material cost and the last factor is the machine investment costs, which include items related to housing, using, and maintaining the additive manufacturing system. Among other things, this includes energy costs, machine purchase, and associated labor costs to operate the system. The base model has a build rate of 6.3 ccm/hr, a utilization of 4500 h/yr, a material cost of 89 €, and a machine investment cost of 500 000 €. For comparison, the base case is shown four times in the figure, with each one shown with a star. On average, the machine costs accounted for 62.9% of the cost estimates in Figure 3.3 (note that the base case is only counted once in the average). This cost was the largest even when building rate was more than tripled and other factors were held constant. This cost was largest in all but one case, where material costs were increased to 600 €/kg. The second largest cost is the materials, which, on average, accounted for 18.0% of the costs; however, it is important to note that this cost is likely to decrease as more suppliers enter the field.[27] Post processing, preparation, oven heating, and building process fix were approximately 8.4%, 5.4%, 3.3%, and 1.9%, respectively.

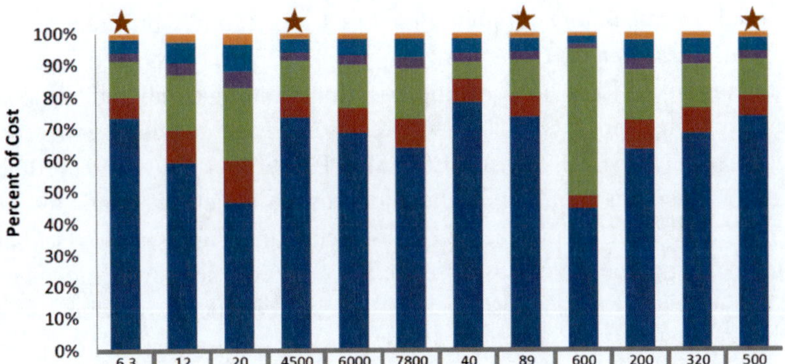

	Building rate in ccm/h: 6.3	12	20	Machine Utilization h/yr: 4500	6000	7800	Material Costs (€/kg): 40	89	600	Machine investment costs (€ thousands): 200	320	500
Building process fix	2%	2%	3%	2%	2%	2%	2%	2%	1%	2%	2%	2%
Preparation costs	4%	7%	9%	4%	5%	6%	5%	4%	3%	6%	5%	4%
Oven	3%	4%	5%	3%	3%	4%	3%	3%	2%	4%	3%	3%
Material costs	11%	17%	23%	11%	14%	16%	5%	11%	46%	16%	14%	11%
postprocessing costs	7%	10%	13%	7%	8%	9%	7%	7%	4%	9%	8%	7%
Machine costs	74%	59%	47%	74%	69%	64%	78%	74%	45%	63%	68%	74%

Source: Lindemann C., U. Jahnke, M. Moi, and R. Koch. "Analyzing Product Lifecycle Costs for a Better Understanding of Cost Drivers in Additive Manufacturing." Proceedings of the 2012 Solid Freeform Fabrication Symposium. <http://utwired.engr.utexas.edu/lff/symposium/proceedingsArchive/pubs/Manuscripts/2012/2012-12-Lindemann.pdf>

Note: The orange star indicates the base model.

Figure 3.3. Cost Distribution of Additive Manufacturing of Metal Parts by varying Factors.

The material costs for additive manufacturing are significant; however, technologies can often be complementary, where two technologies are adopted alongside each other and the benefits are greater than if they were adopted individually. One example is computer aided design and computer aided manufacturing, as both are needed to be utilized for the other to be valuable. Additive manufacturing and the raw materials that are used may be a condition where they are complementary.[28] All additive manufacturing requires raw materials, and according to Stoneman (2002) this may create a feedback loop.[29]

Increasing adoption of additive manufacturing may lead to a reduction in raw material cost through economies of scale. The reduced cost in raw material might then propagate further adoption of additive manufacturing. There may also be economies of scale in raw material costs if particular materials become more common rather than a plethora of different materials.

Plastic Material Costs: Atzeni et al. (2010) compared the costs of manufacturing a lamp holder using injection molding compared to the additive manufacturing process of selective laser sintering using two different machines: EOS SLS P730 and EOS SLS P390.[30] A significant portion of the cost for injection molding is the mold itself, which accounts for between 84.6% and 97.7% of the cost as seen in Figure 3.4. For additive manufacturing, the major costs are the machine cost per part, which is between 58.7% and 65.9% of the cost, and the material cost per part, which is between 29.1% and 30.4% of the cost. The P730 is cost effective for production volumes of 73 000 or less while the P390 is cost effective for 87 000 or less.

Hopkinson and Dickens (2003) also investigate the additive manufacturing costs of a polymer part, as discussed in Section 4.[31] The costs are calculated for two parts, a lever and a cover, using stereolithography, fused deposition modeling, and laser sintering. A cost breakout for the lever is provided in Figure 3.5 and Table 3.1. The material cost represented 25% of the cost for stereolithography, 39% for fused deposition modeling, and 74% for laser sintering. Ruffo et al. (2006a) conduct a similar analysis using the same part.[32] The cost of additive manufactured parts is calculated by Ruffo et al. using an activity based cost model, where each cost is associated with a particular activity. They make an estimate that compares with Hopkinson and Dickens and another estimate that uses recycling of material. As illustrated in Figure 3.6, material is 69% of the cost in the first estimate and 55% in the second estimate.

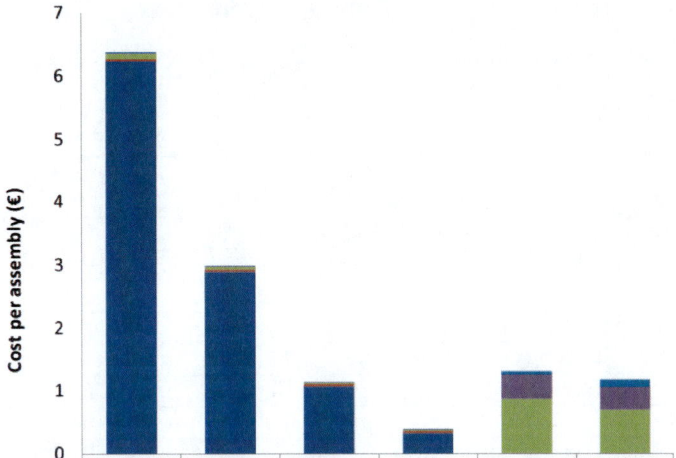

	IM 5000	IM 20000	IM 100000	IM 500000	AM P730	AM P390
Machine operator cost per part	0.018	0.009	0.004	0.001	0.054	0.117
Material cost per part	0.011	0.011	0.011	0.011	0.382	0.360
Machine cost per part	0.083	0.042	0.024	0.014	0.867	0.694
Assembly cost per part	0.035	0.035	0.035	0.035	0.012	0.012
Mold cost per part	6.240	2.900	1.075	0.335		

Note: The number following IM is the number of assemblies; thus, IM 5000 is injection molding with 5000 assemblies made. The number following AM is the model of the machine; thus, AM P730 is additive manufacturing machine EOS SLS P730. P390 is the EOS SLS P390.

Figure 3.4. Cost Comparison of Injection Molding and Additive Manufacturing for a Selected Product, Atzeni et al. (2010).

Table 3.1. Cost Breakout, Hopkinson and Dickens (2003)

	Stereolithography	Fused deposition modeling	Laser sintering
Number per platform	190	75	1056
Platform build time	26.8	67.27	59.78
Production rate per hour	7.09	1.11	17.66
Hours per year in operation	7 884	7 884	7 884
production volume total per year	55 894	8 790	139 269
Machine and ancillary equipment (€)	1 040 000	101 280	340 000
Equipment depreciation cost per year (€)	130 000	12 660	42 500

Table 3.1. (Continued)

	Stereolithography	Fused deposition modeling	Laser sintering
Machine maintenance cost per year (€)	89 000	10 560	30 450
Total machine cost per year (€)	219 000	23 220	72 950
Machine cost per part (€)	3.92	2.64	0.52
Machine operator cost per hour (€)	5.30	5.30	5.30
Set-up time to control machine (min)	33	10	120
Post-processing time per build (min)	49	60	360
Labor cost per build (€)	7.24	6.18	42.37
Labor cost per part (€)	0.04	0.08	0.04
Material per part (kg)		0.0035	
Support material per part (kg)	0.0047	0.0016	
Build material cost per kg (€)		400.00	
Support material cost per kg (€)	275.20	216.00	54.00
Cost of material used in one build (€)			1 725.72
Material cost per part (€)	1.29	1.75	1.63
Total cost per part (€)	5.25	4.47	2.20

3.2.2. Machine Cost

In addition to material costs, machine cost is one of the most significant costs involved in additive manufacturing. The average selling price of an industrial additive manufacturing system was $73 220 in 2011.[33] Although the price is up from $62 570 in 2010, the price has fallen for most years prior to this point. Between 2001 and 2011, the price decreased 51% after adjusting for inflation.[34] While the trends in machine costs are generally downward, large differences remain between the costs for polymer-based systems and metal-based systems, and the tremendous growth in sales of low-cost, polymer-based systems during this time has strongly influenced the average selling price of additive manufacturing systems.

For metal material cost studies, Hopkinson and Dickens (2003) showed that machine costs ranged from 23% to 75% of a metal part, as seen in Table 3.1. The cost difference between the different types of additive manufacturing machinery was quite significant ranging between $0.1 million typically for polymer systems and $1.0 million typically for metal systems. One might surmise that the proportion might have decreased over time; however, the machine cost estimates for Lindemann et al. (2012) ranged from 45% to 78%

of the cost of a metal part, as seen in Figure 3.3. Atzeni et al. (2010) show that machine cost per part was between 59% and 66% of the cost of a plastic part, as seen in Figure 3.4.

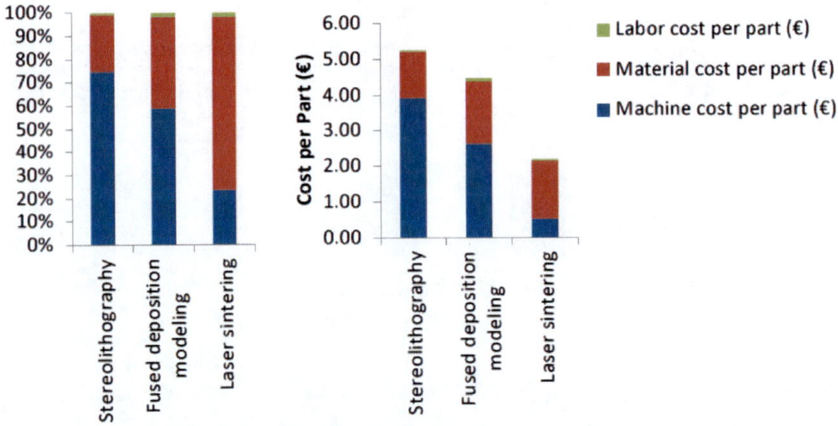

Figure 3.5. Cost Breakout, Hopkinson and Dickens (2003).

3.2.3. Build Envelope and Envelope Utilization

The size of the build envelope[35] and the utilization of this envelop both have an impact on the cost of an additive manufactured product. The size of the build envelope has two impacts. First, products can only be built to the size of the build envelope, which means that it might not be possible to build some products using additive manufacturing technologies without enlarging the build envelope. The second impact of the build envelope is related to utilizing the total amount of build capacity. A significant efficiency factor lies in the ability to exhaust the available build space. For example, Baumers et al. (2011) examined the impact of capacity utilization on energy using six different machines (Arcam - A1, MTT Group - SLM 250, EOS GmbH - EOSINT M 270, Concept Laser GmbH - M3 Linear, Stratasys Inc - FDM 400 mc, and EOS GmbH - EOSINT P390) and four different materials (titanium, stainless steel, and two kinds of polymers). As seen in Figure 3.7, the full build case, where the build envelope is fully utilized, uses less energy per kilogram deposited than one single part being produced for all six different machines. The EOSINT P 390 has the largest build volume and has the largest difference in energy consumption between a single part and full build.

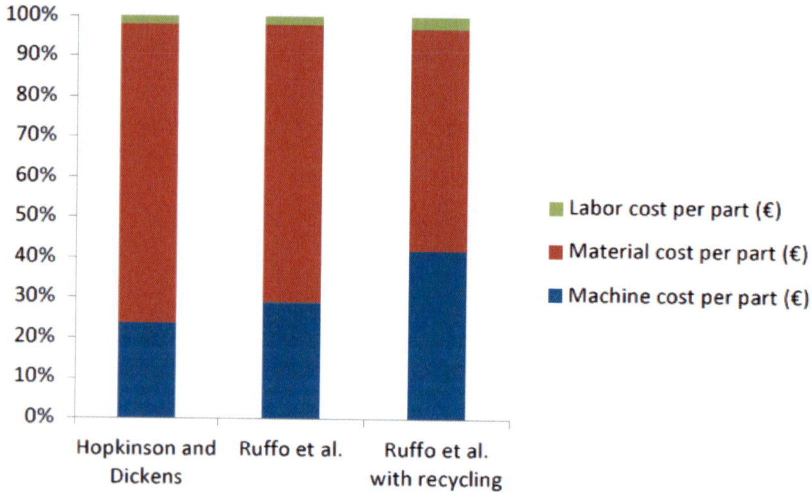

Figure 3.6. Cost Comparison for Selective Laser Sintering.

3.2.4. Build Time

Build time is a significant component in regard to estimating the cost of additive manufacturing and a number of software packages are available for estimating build time.[36,37] There tends to be two approaches to estimating build time: 1) detailed analysis and 2) parametric analysis.[38] Detailed analysis utilizes knowledge about the inner workings of a system, while parametric analysis utilizes information on process time and characteristics such as layer thickness. Build time estimations tend to be specific to the system and material being used. Although this is an important factor in the cost of additive manufacturing, the details of build time are beyond the scope of this report.

3.2.5. Energy Consumption

Some cost studies for additive manufacturing, such as Hopkinson and Dickens (2003), included an examination of energy consumption, but they did not include energy in their reporting, as it contributed less than one percent to the final cost.[39] Energy consumption, however, is an important factor in considering the cost of additive manufacturing compared to other methods of manufacturing, especially in terms of examining the costs from cradle to grave. Energy studies on additive manufacturing, however, tend to focus only on the energy used in material refining and by the additive manufacturing system itself. These studies are discussed below.

Metal: As discussed previously, Baumers et al. (2011) examined energy consumption among a number of machines.[40] The results shown in Figure 3.7

provide the results for energy consumption among these machines. Morrow et al. (2007) compares direct metal deposition to conventional tool and die manufacturing.[41] This work identifies that energy consumption is driven by the solid-to-cavity volume ratio. At low ratios, the additive manufacturing process of direct metal deposition minimizes energy, while at high ratios computer numeric controlled milling minimizes energy consumption. Other studies tend to focus on accurately predicting energy consumption and minimizing energy consumption for additive manufacturing. Envelope utilization and build orientation are among the issues for reducing energy consumption. Mognol, Lepicart, and Perry (2006) examine the impact of part orientation for three systems: Stratasys FDM 3000, 3D Systems Thermojet, and EOS EOSINT M250 Xtended.[42] They examined 18 positions for a single part. Due to the change in the position of the part, the energy consumed could increase between 75% and 160% depending on the system, as illustrated in Figure 3.8. This figure also illustrates that the position for one system may have low energy consumption, but for another system it might not have a low consumption.

Plastic Material: Telenko and Seepersad (2012) examined energy consumed in the production of nylon parts using selective laser sintering and compared these results to that of injection molding.[43] This analysis included a small build of 50 parts and a full build of 150 parts. The results are displayed in Figure 3.9 with injection mold values (IM) being shown both with the energy consumed for the production of the mold and without the mold. As seen in the figure, the small build for selective laser sintering used less energy than the small build for injection molding (including the energy for the mold). However, the energy for the full build was approximately 69% higher. For the full build, approximately 60% of the energy was used in nylon production and 37% was used in part manufacture for selective laser sintering.

Sreenivasan and Bourell (2009) examined the energy use of selective laser sintering using nylon material, building two “full chamber build[s]” of prosthetic parts.[44] They identify the components that are major consumers of energy: chamber heaters (37%), stepper motors for piston control (26%), roller drives (16%), and the laser (16%).

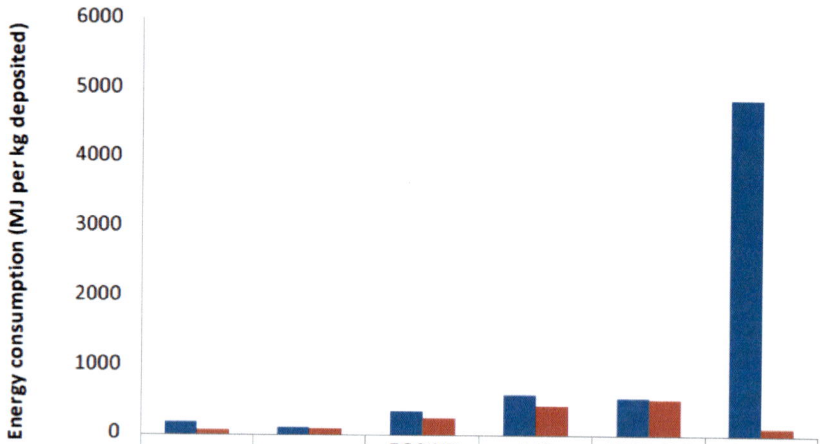

	A1 (Metal)	SLM250 (Metal)	EOSINT M 270 (Metal)	M3 Linear (Metal)	FDM 400 mc (Polymer)	EOSINT P 390 (Polymer)
Single Part	177	106	339	588	536	4849
Full Build	61	83	241	423	519	107

Figure 3.7. Energy Consumption per kg Deposited (Baumers et al. 2011).

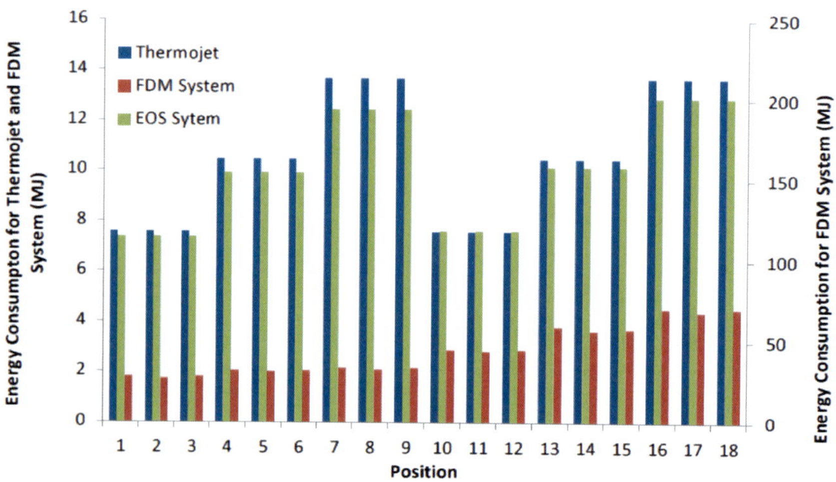

Figure 3.8. Energy Consumption, Magnol, Lepicart, and Perry (2006).

3.2.6. Labor

As illustrated in Figure 3.5 and Figure 3.6, labor tends to be a small portion of the additive manufacturing cost. Labor might include removing the finished product or refilling the raw material among other things. From Figure 3.6, Hopkinson and Dickens estimate labor at 2% of the cost, while Ruffo et al. estimate it at 2% and 3%. It is important to note that additional labor is built into the other costs such as the material cost and machine cost, as these items also require labor to produce.

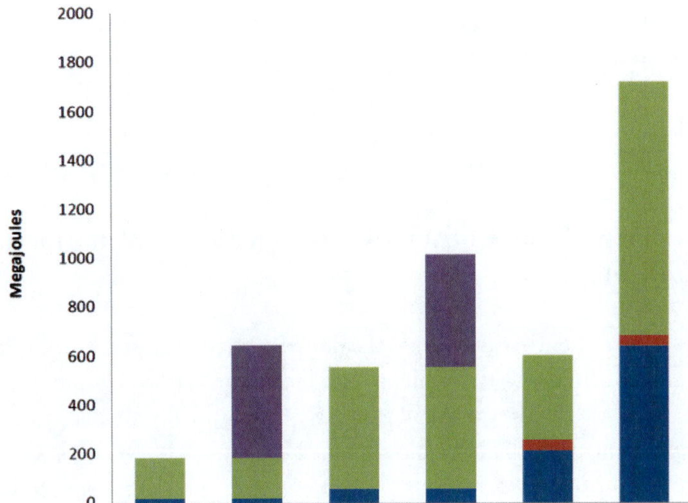

	IM Small Build	IM Small Build (incl. mold)	IM Full Build	IM Full Build (incl. mold)	SLS Small Build	SLS Full Build
Recycled Steel Mold Production		461		461		
Nylon Production	165	165	496	496	345	1036
Build Preheat					43	43
Part Manufacture	20	20	59	59	214	641

Figure 3.9. Energy Efficiency of Selective Laser Sintering, Cassandra and Seepersad (2012), megajoules.

3.3. Product Enhancements and Quality

Although the focus of this report is the costs of additive manufacturing, it is important to note that there are product enhancements and quality differences due to using this technology. There is more geometric freedom with additive manufacturing and it creates more flexibility; however, there are limitations, as some designs require support structures and means for dissipating heat in production.[45] However, complexity does not increase the cost of production as it does with traditional methods. With the exception of the design cost, each product produced can be customized at little or no expense. There is significant need for custom products in the medical sector for replacement joint implants, dental work, and hearing aids among other things.[46] There is also the possibility of customers designing their own products or customizing them. One concern with additive manufacturing, however, is quality assurance. Currently, there is a need for standard methods to evaluate and ensure accuracy, surface finish, and feature detail to achieve desired part quality.

4. Cost Models and Comparisons

4.1. Two Major Contributions to Additive Manufacturing Cost Modeling

There are two cost models that receive significant attention in additive manufacturing: 1) Hopkinson and Dickens (2003) and 2) Ruffo et al. (2006a).[47,48,49] The cost of additive manufactured parts is calculated by Hopkinson and Dickens based on calculating the average cost per part and three additional assumptions: 1) the system produces a single type of part for one year, 2) it utilizes maximum volumes, and 3) the machine operates for 90% of the time. The analysis includes labor, material, and machine costs. Other factors such as power consumption and space rental were considered but contributed less than one percent of the costs; therefore, they were not included in the results. The average part cost is calculated by dividing the total cost by the total number of parts manufactured in a year. Costs can be broken into machine costs, labor costs, and material costs. The costs are calculated for two parts, a lever and a cover, using stereolithography, fused deposition modeling, and laser sintering. A cost breakout for the lever is provided in Figure 3.5 and Table 3.1, which shows that in this analysis laser sintering was the cheapest additive manufacturing process for this product. Machine cost was the major contributing cost factor for stereolithography and fused deposition modeling while the material cost was the major contributor for laser sintering.

Hopkinson and Dickens estimate an annual machine cost per part where the machine completely depreciates after eight years; that is, it is the sum of depreciation cost per year (calculated as machine and ancillary equipment divided by 8) and machine maintenance cost per year divided by production volume. The result is a machine cost per part that is constant over time, as seen in Figure 4.1. Also seen in the figure is a comparison to injection molding.

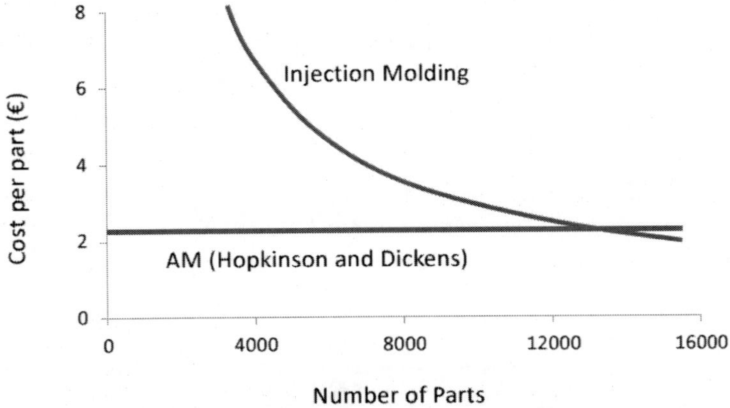

Adapted from Hopkinson and Dickens (2003).

Figure 4.1. Hopkinson and Dickens (2003) Cost Model Compared to Injection Molding.

The cost of additive manufactured parts is calculated by Ruffo et al. using an activity based cost model, where each cost is associated with a particular activity. They produce the same lever that Hopkinson and Dickens produced using selective laser sintering. In their model, the total cost of a build (C), is the sum of raw material costs and indirect costs. The raw material costs are the price (Pmaterial), measured in euros per kilogram, multiplied by the mass in kg (M). The indirect costs are calculated as the total build time (T) multiplied by a cost rate (Pindirect). The total cost of a build is then represented as:

$$C = P_{material} * M + P_{indirect} * T$$

The cost per part is calculated as the total cost of a build (C) divided by the number of parts in the build. In contrast, Ruffo et al. indicate that the time and material used are the main variables in the costing model. It was assumed that the machine worked 100 hours/week for 50 weeks/year (57% utilization). The estimated indirect cost per hour is shown in Table 4.1. Their cost model and the total costs are shown in Figure 4.2.

There are three different times that are calculated in Ruffo et al.'s model: 1) "time to laser scan the section and its border in order to sinter;" 2) "time to add layers of powder;" and 3) "time to heat the bed before scanning and to cool down slowly after scanning, adding layers of powder or just waiting time to reach the correct temperature." The sum of these times is the build time (*T*) and the resulting cost model along with the Hopkinson and Dickens model is shown in Figure 4.3. The Ruffo et al. model has a jagged saw tooth shape to it, which is due to the impact of a new line, layer, or build. Each time one of these is added, average costs increase irregularly from raw material consumption and process time. At 1600 parts, the cost of the lever is estimated at €2.76 per part compared to Hopkinson and Dickens €2.20 for laser sintering. Ruffo et al. also conducted an examination where unused material was recycled. In this examination, the per-unit cost was € 1.86. A comparison of the costs is made in Figure 3.6.

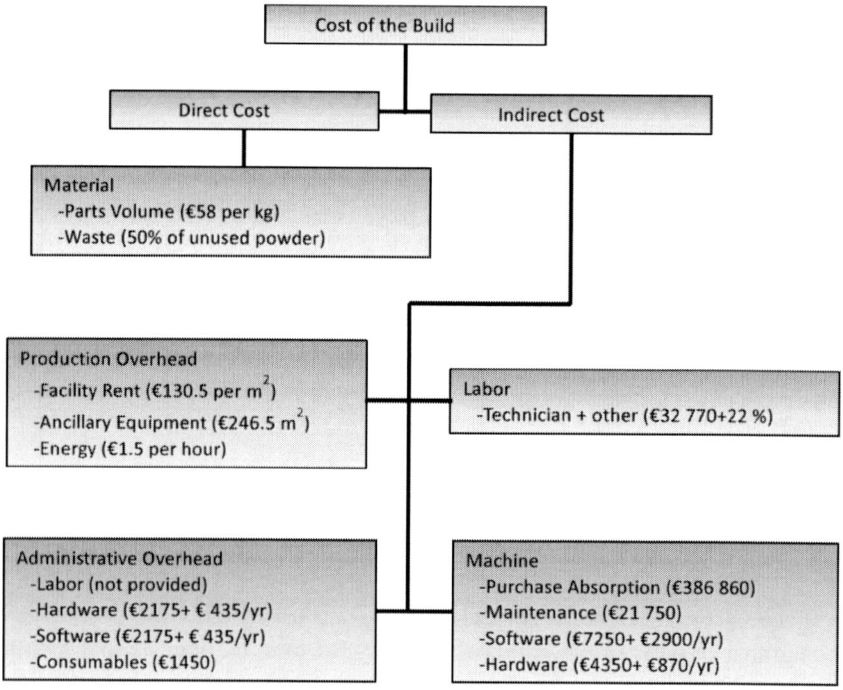

Figure 4.2. Ruffo, Tuck, and Hague Cost Model.

Table 4.1. Indirect Cost Activities (Ruffo, Tuck, and Hague 2006a)

Activity	Cost/hr (€)
Production labor/machine hour	7.99
Machine costs	14.78
Production overhead	5.90
Administrative overhead	0.41

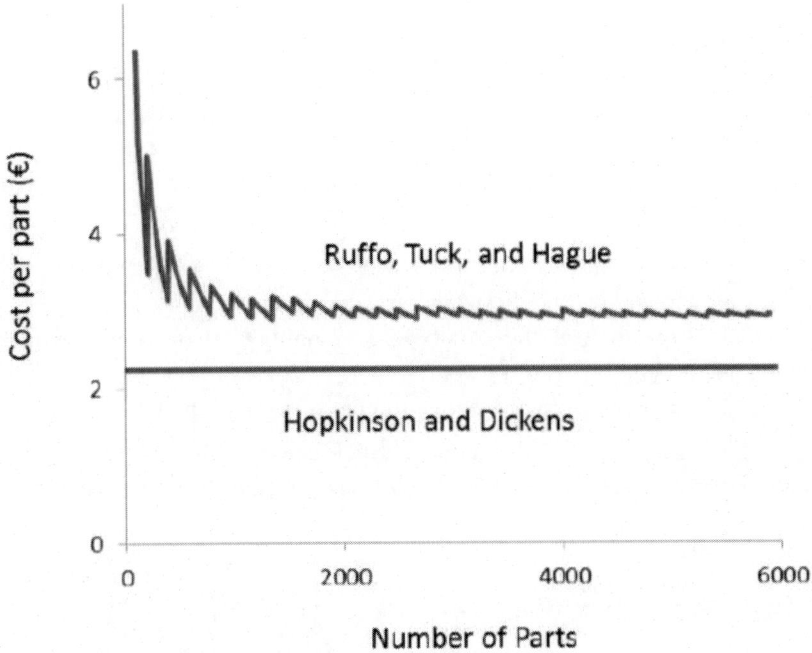

Adapted from Ruffo et al. and Hopkinson and Dickens.

Figure 4.3. Cost Model Comparison (Ruffo, Tuck, and Hague vs. Hopkinson and Dickens).

Many of the cost studies assume a scenario where one part is produced repeatedly; however, one of the benefits of additive manufacturing is the ability to produce different components simultaneously. Therefore, a "smart mix" of components in the same build might achieve reduced costs. In a single part production, the per part cost for a build is the total cost divided by the number of parts; however, the cost for different parts being built simultaneously is more complicated. Ruffo and Hague (2007) compare three costing methodologies for assessing this cost. The first method is based on parts volume where

$$Cost_{P_i} = \left(\frac{V_{P_i}}{V_B} \right) * Cost_B$$

Where

$Cost_{P_i}$= cost of part i

V_{P_i}= volume of part i

V_B= volume of the entire build

$$Cost_B = \sum \frac{indirect\_costs}{working\_time} (t_{xy} + t_z + t_{HC}) + \frac{direct\_cost}{mass\_unit} m_B$$

m_B = mass of the planned production proportional to the object volumes, and the time to manufacturing the entire build

t_{xy} = time to laser-scan the section and its border to sinter powder

t_z = time to add layers of powder

t_{HC} = time to heat the bed before scanning and to cool down after scanning and adding layers of powder

i = an index going from one to the number of parts in the build

$Cost_B$ also equals *C* from above, which is the total cost of a build. The second method is based on the cost of building a single part and is represented as the following:

$$Cost_{P_i} = \frac{\gamma_i * Cost_B}{n_i}$$

where

$$\gamma_i = \frac{Cost^*_{P_i} + n_i}{\sum_j (Cost^*_{P_j} * n_j)}$$

Also, *i* is the index of the part being calculated, *j* is the index for all parts manufactured in the same bed, n_i is the number of parts identified with i, and $Cost^*_{P_i}$ is the cost of a single part *i* estimated using the earlier equation for *C*. The third method is based on the cost of a part built in high-volume. It is similar to the second method, only the cost variables in γ_i are calculated using a high number of parts rather than a single part. It is represented as the following:

$$Cost_{P_i} = \frac{\gamma_i^{\infty} * Cost_B}{n_i}$$

where

$$\gamma_i^{\infty} = \frac{Cost_{P_i}^{\infty} + n_i}{\Sigma_j \left(Cost_{P_j}^{\infty} * n_j \right)}$$

Where $Cost_{P_i}^{\infty}$ is a hypothetical number, which approaches infinity, of manufactured parts *i*.

Ruffo and Hague use a case study to evaluate the validity of estimating the per part cost. The results suggest that only the third model provides a "fair assignment method." The other two were identified as being inappropriate due to the result drastically reducing the estimated cost of larger components at the expense of smaller parts.

4.2. Other Comparisons to Traditional Manufacturing

Atzeni and Salmi (2011) showed that the per assembly processing cost for a landing gear assembly for a 1:5 scale model of the P180 Avant II by Piaggio Aero Industries S.p.A. (i.e., the machine cost per assembly), with an estimated five years of useful life, was €472.50 for the additive manufacturing process of selective laser sintering (see Table 4.2). Compared to high-pressure die-casting, the mold cost and processing cost per part were €0.26 + €21 000/N, where N is the number of parts produced. For production runs of less than 42, selective laser sintering was more cost effective than the traditional process of high-pressure die-casting (see Figure 4.4).

The aerospace industry often uses costly raw materials, which have high performance and low weight. These high performance materials are not only costly to purchase, but can also be costly to machine down using traditional manufacturing methods. Allen (2006) compares additive manufacturing to machining for aero engine parts.50 This work provides a more generic comparison of the two processes. The cost of providing a "near net shape" using machining was estimated as the following:

Table 4.2. Production Costs Compared, Atzeni and Salmi (2011)

	High-Pressure Die-Cast Part (€)	Selective Laser Sintering Part (€)
Material cost per part	2.59	25.81
Pre-processing cost per part	-	8.00
Processing cost per part*	0.26+21 000/N	472.50
Post-processing cost per part	17.90	20.00
Assembly	0.54	-
Total	**21.29+21 000/N**	**526.31**

*Includes the mold for die-casting.

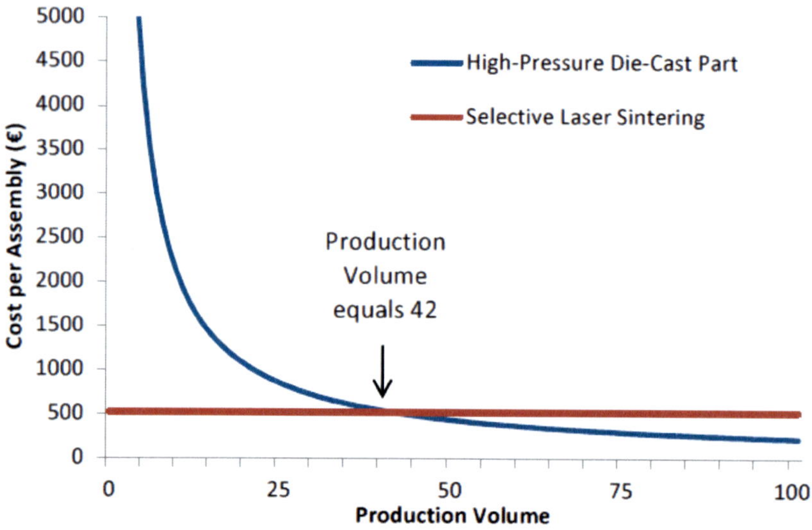

Figure 4.4. Breakeven Point for High-Pressure Die-Casting and Selective Laser Sintering, Atzeni and Salmi (2011).

$$C_s = (V * \rho * C_f) * (V - v) * \rho * C_m$$

Where

C_s= cost of providing a "near net shape" using machining

V = volume of original billet

ρ = density of titanium

C_f = cost of ring rolled forged material
v = volume of component
C_m = cost of machining

The cost of producing a "near net shape" using additive manufacturing was estimated as the following:

$$C_a = v * \rho * C_d$$

Where
C_a= cost of producing a "near net shape" using additive manufacturing
v = volume of component
ρ = density of titanium
C_d = specific cost of deposited titanium

This work concluded that additive manufacturing is cost effective in instances where the buy/fly ratio is 12-1 compared to more "conventional" ratios which tend to be lower.

Note that the buy/fly ratio is calculated as the volume of the billet (V) divided by the volume of the component (v). It is a means for representing how much material must be machined away. Allen concludes that additive manufacturing techniques are attractive for components with a high buy/fly ratio, have a complex shape that requires significant machining, has a high material cost, and has slow machining rates.

4.3. Additive Manufacturing Cost Advantage

Many of the cost studies examine costs such as material and machine costs; however, many of the benefits may be hidden in inventory and supply chain costs. For instance, a dollar invested in automotive assembly takes 10.9 days to return in revenue. It spends 7.9 days in material inventory, waiting to be utilized. It spends 19.8 hours in production time and another 20.6 hours in down time when the factory is closed. Another 1.3 days is spent in finished goods inventory.51 Moreover, of the total time used, only 8% is spent in actual production. According to concepts from lean manufacturing, inventory and waiting, which constitute 92% of the automotive assembly time, are two of seven categories of waste. This is just the assembly of an automobile. The production of the engine parts, steering, suspension, power train, body, and others often occur separately and also have inventories of their own. Additionally, all of these parts are transported between locations. The average shipment of manufactured transportation equipment travels 801 miles. For the US, this amounts to 45.3 billion ton-miles of transportation equipment being moved annually. Because additive manufacturing can, in some instances now and possibly more in the future, build an entire assembly in one build, it reduces the need for some of the transportation and inventory costs, resulting in impacts throughout the supply chain. It is important to note that the ability to produce more complex assemblies, such as those in an automobile, is still developing and involves some speculation about future capabilities. In addition to building complete or partial assemblies, there is also the potential of reducing the size of the supply chain through distributed manufacturing. Therefore, in order to understand the cost difference between additive manufacturing and other processes, it is necessary to examine the costs from raw material extraction to production and through the sale of the final product. This might be represented as:

$$C_{AM} = (MI_{R,AM} + MI_{M,AM}) + (P_{E,AM} + P_{R,AM} + P_{M,AM}) + (FGI_{E,AM} + FGI_{R,AM} + FGI_{M,AM}) + WT_{AM} + RT_{AM} + T_{AM}$$

Where

C_{AM} = Cost of producing an additive manufactured product

MI = Cost of material inventory for refining raw materials (R) and for manufacturing (M) for additive manufacturing (AM)

P = Cost of the process of material extraction (E), refining raw materials (R), and manufacturing (M), including administrative costs, machine costs, and other relevant costs for additive manufacturing (AM)

FGI = Cost of finished goods inventory for material extraction (E), refining raw materials (R), and manufacturing (M) for additive manufacturing (AM)

WT_{AM} = Cost of wholesale trade for additive manufacturing (AM)

RT_{AM}= Cost of retail trade for additive manufacturing (AM)

T_{AM} = Transportation cost throughout the supply chain for an additive manufactured Product (AM)

This could be compared to the cost of traditional manufacturing, which could be represented as the following:

$$C_{Trad} = (MI_{R,Trad} + MI_{I,Trad} + MI_{A,Trad}\big) + \big(P_{E,Trad} + P_{R,Trad} + P_{I,Trad} + P_{A,Trad}) + \big(FGI_{E,Trad} + FGI_{R,Trad} + FGI_{I,Trad} + FGI_{A,Trad}\big) + WT_{Trad} + RT_{Trad} + T_{Trad}$$

Where

C_{Trad} = Cost of producing a product using traditional processes (*Trad*)

MI = Cost of material inventory for refining raw materials (*R*), producing intermediate goods (*I*), and assembly (*A*) for traditional manufacturing (*Trad*)

P = Cost of the process of material extraction (*E*), refining raw materials (*R*), producing intermediate goods (*I*), and assembly (*A*), including administrative costs, machine costs, and other relevant costs for traditional manufacturing (*Trad*)

FGI = Cost of finished goods inventory for material extraction (*E*), refining raw materials (*R*), producing intermediate goods (*I*), and assembly (*A*) for traditional manufacturing (*Trad*)

WT_{Trad} = Cost of wholesale trade for traditional manufacturing (*Trad*)

RT_{Trad} = Cost of retail trade for traditional manufacturing (*Trad*)

T_{Trad} = Transportation costs throughout the supply chain for a product made using traditional manufacturing (*Trad*)

Currently, there is a better understanding about the cost of the additive manufacturing process cost (P_{AM}) than there is for the other costs of additive manufacturing. Additionally, most cost studies examine a single part or component; however, it is in the final product where additive manufacturing might have significant cost savings. Traditional manufacturing requires numerous intermediate products that are transported and assembled, whereas additive manufacturing can achieve the same final product with fewer component parts and multiple components built either simultaneously or in the same location. For example, consider the future possibility of an entire jet engine housing being made in one build using additive manufacturing compared to an engine housing that has parts made and shipped for assembly from different locations with each location having its own factory, material inventory, finished goods inventory, administrative staff, and transportation infrastructure among other things. Additionally, the jet engine housing might be made using less material, perform more efficiently, and last longer because the design is not limited to the methods used in traditional manufacturing; however, many of these benefits would not be captured in the previously mentioned cost model. To capture these benefits one would need to include a cradle to grave analysis.

4.4. Additive Manufacturing Total Advantage

At the company level, the goal is to maximize profit; however, at the societal level there are multiple stakeholders to consider and different costs and benefits. At this level, one might consider the goal to be to minimize resource use and maximize utility. Dollar values are affected by numerous factors such as scarcity, regulations, and education costs among other things that impact how resources are efficiently allocated. The allocation of resources is an important issue; however, understanding the societal impact of additive manufacturing requires separating resource allocation issues from resource utilization issues. The factors of production are, typically, considered to be land (i.e., natural resources), labor, capital, and entrepreneurship; however, capital includes machinery and tools, which themselves are made of land and labor. Additionally, a major element in the production of all goods and services is time, as illustrated in many operations management discussions. Therefore, one might consider the most basic elements of production to be land, labor, human capital, entrepreneurship, and time. The human capital and entrepreneurship utilized in producing additive manufactured goods are important, but these are complex issues that are not a focus of this report. The remaining items land, labor, and time constitute the primary cost elements for production. It is important to note that there is a tradeoff between time and labor (measured in labor hours per hour), as illustrated in Figure 4.5. For example, it takes one hundred people less time to build a house than it takes for one person to build a house. It is also important to note that there is also a tradeoff between time/labor and land (i.e., natural resources), as illustrated in Figure 4.6. For example, a machine can reduce both the time and the number of people needed for production, but utilizes more energy. The triangular plane in the figure represents possible combinations of land, labor, and time needed for producing a manufactured good. Moving anywhere along this plane is simply an alteration of resource use. A company can maximize profit by either altering resources or by reducing the resources needed for production. Moving along the plane in Figure 4.6 may result in a more efficient allocation of resources for a firm and for society; however, it does not reduce the combination of resources needed for production. Therefore, when examining the cost and benefits of a product or process from a societal perspective, it becomes apparent that one needs to measure land, labor, and time needed for production in order to understand whether there has been a reduction in the combination of resources needed to produce a manufactured good. If additive manufacturing results in a reduction in the resources needed for production, then that plane will move toward the origin as illustrated in Figure 4.6.

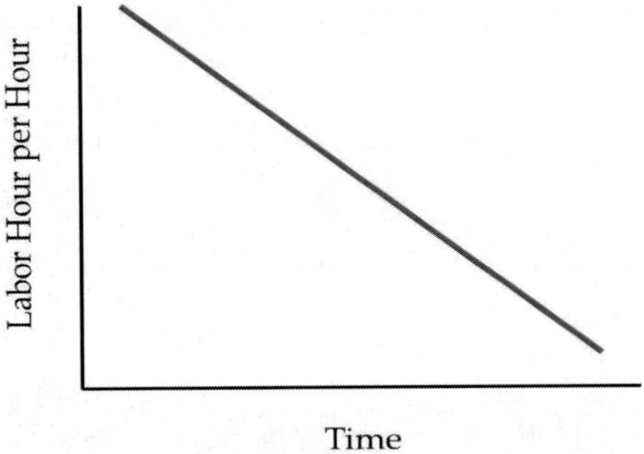

Figure 4.5. Time and Labor Needed to Produce a Manufactured Product.

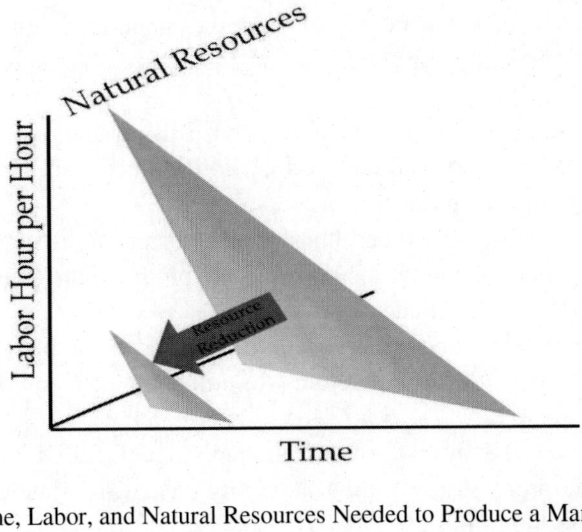

Figure 4.6. Time, Labor, and Natural Resources Needed to Produce a Manufactured Product.

In addition to production, manufactured goods are produced to serve a designated purpose. For example, automobiles transport objects and people; cell phones facilitate communication; and monitors display information. Each item produced is designed for some purpose. In the process of fulfilling this purpose more resources are expended in the form of land, labor, and time. Additionally, a product with a short life span results in more resources being expended to reproduce the product. Additionally, the disposal of the old product may result in expending further resources. Additive manufactured products may provide product enhancements, new abilities, or an extended useful life. The total advantage of an additive manufactured good is the difference in the use of land, labor, and time expended on production, utilization, and disposal combined with the utility gained from the product compared to that of traditional manufacturing methods. This can be represented as the following:

$$TA_L = \left(L_{AM,P} + L_{AM,U} + L_{AM,D}\right) - \left(L_{T,P} + L_{T,U} + L_{T,D}\right)$$

$$TA_{LB} = (\ LB_{AM,P} + LB_{AM,U} + LB_{AM,D}) - (\ LB_{T,P} + LB_{T,U} + LB_{T,D})$$

$$TA_T = \left(T_{AM,P} + T_{AM,U} + T_{AM,D}\right) - \left(T_{T,P} + T_{T,U} + T_{T,D}\right)$$

$$TA_U = U(P_{AM}) - U(P_T)$$

TA = The total advantage of additive manufacturing compared to traditional methods for Land *(L)*, labor *(LB)*, time *(T)*, and utility of the product *(U)*.

L = The land or natural resources needed using additive manufacturing processes *(AM)* or traditional methods *(T)* for production *(P)*, utilization *(U)*, and disposal *(D)* of the product

LB = The labor hours per hour needed using additive manufacturing processes *(AM)* or traditional methods *(T)* for production *(P)*, utilization *(U)*, and disposal (D) of the product

T = The time needed using additive manufacturing processes *(AM)* or traditional methods *(T)* for production *(P)*, utilization *(U)*, and disposal *(D)* of the product

$U(P_{AM})$ = The utility of a product manufactured using additive manufacturing processes, including the utility gained from increased abilities, enhancements, and useful life.

$U(P_T)$ = The utility of a product manufactured using traditional processes, including the utility gained from increased abilities, enhancements, and useful life. In this case production includes material extraction, material refining, manufacturing, and transportation among other things. Unfortunately, our current abilities fall short of being able to measure all of these items for all products; however, it is important to remember that these items must be considered when measuring the total advantage of additive manufacturing. An additional challenge is that land, labor, time, and utility are measured in different units, making them difficult to compare. An additive manufactured product might require more labor but reduce the natural resources needed. In this instance, there is a tradeoff.

5. Implementation and Adoption of Additive Manufacturing

Additive manufacturing is significantly different from traditional methods; thus, determining when and how to take advantage of the benefits of additive manufacturing is a challenge in and of itself. Additionally, the manufacturing industry is oriented toward optimizing production using traditional methods. Identifying products that benefit from increased complexity, or being produced in closer proximity to consumers, or understanding the impact on inventory is complex and difficult as it impacts factors that are difficult to measure.

5.1. Additive Manufacturing and Firm Capabilities

In order to create products and services, a firm needs resources, established processes, and capabilities.[52] Resources include natural resources, labor, and other items needed for production. A firm must have access to resources in order to produce goods and services. The firm must also have processes in place that transform resources into products and services. Two firms may have the same resources and processes in place; however, their products may not be equivalent due to quality, performance, or cost of the product or service. This difference is due to the capabilities of the firm; that is, capabilities are the firm's ability to produce a good or service effectively. Kim and Park (2013) present three entities of capabilities (see Figure 5.1): controllability, flexibility, and integration.[53]

Controllability is the firm's ability to control its processes. Its primary objective is to achieve efficiency that minimizes cost and maximizes accuracy and productivity. Flexibility is the firm's ability to deal with internal and external uncertainties. It includes reacting to changing circumstances while sustaining few impacts in time, cost, or performance. According to Kim and Park, there is a tradeoff between controllability and flexibility; that is, in the short term, a firm chooses combinations of flexibility and controllability, sacrificing one for the other as illustrated in Figure 5.2. Over time, a firm can integrate and increase both flexibility and controllability through technology or knowledge advancement among other things. In addition to the entities of capabilities, there are categories of capabilities or a chain of capabilities, which include basic capabilities, process-level capabilities, system-level capabilities, and performance. As seen in Figure 5.3, basic capabilities include overall knowledge and experience of a firm and its employees, including their engineering skills, safety skills, and work ethics among other things. Process-level capabilities include individual functions such as assembly, welding, and other individual activities. System-level capabilities include bringing capabilities together to transform resources into goods and services. The final item in the chain is performance, which is often measured in profit, revenue, or customer satisfaction among other things.

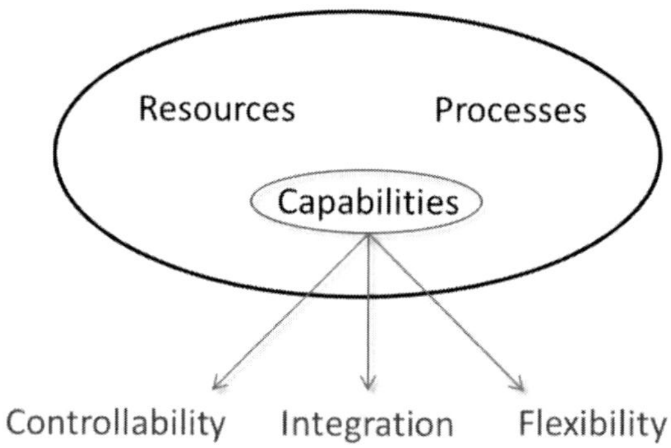

Adapted from Kim, Bowon and Chulsoon Park. (2013). "Firms' Integrating Efforts to Mitigate the Tradeoff Between Controllability and Flexibility." International Journal of Production Research. 51(4): 1258-1278.

Figure 5.1. Necessities of a Firm.

Adopting a new technology, such as additive manufacturing, can have significant impacts on a firm's capabilities. As discussed in the previous sections, in some instances the per unit cost can be higher for additive manufacturing than for traditional methods. The result is that a firm sacrifices controllability for flexibility; thus, it makes sense for those firms that seek a high flexibility position to adopt additive manufacturing. In some instances, however, additive manufacturing can positively affect controllability. Additive manufacturing can reduce costs for products that have complex designs that are costly to manufacture using traditional methods. As the price of material and systems comes down for additive manufacturing, the controllability associated with this technology will increase, making it attractive to more firms.

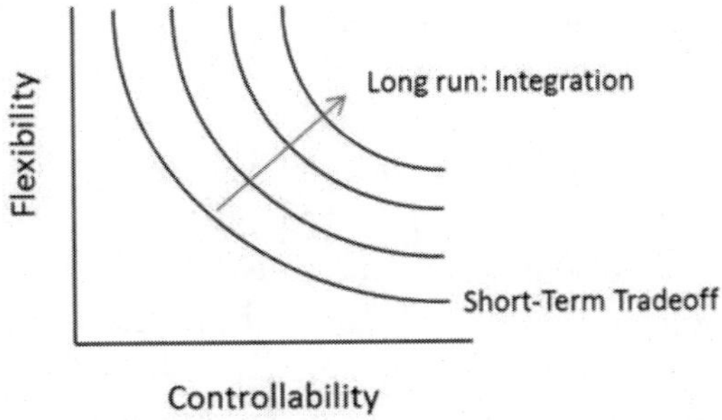

Adapted from Kim, Bowon and Chulsoon Park. (2013). "Firms' Integrating Efforts to Mitigate the Tradeoff Between Controllability and Flexibility." International Journal of Production Research. 51(4): 1258-1278.

Figure 5.2. Flexibility and Controllability.

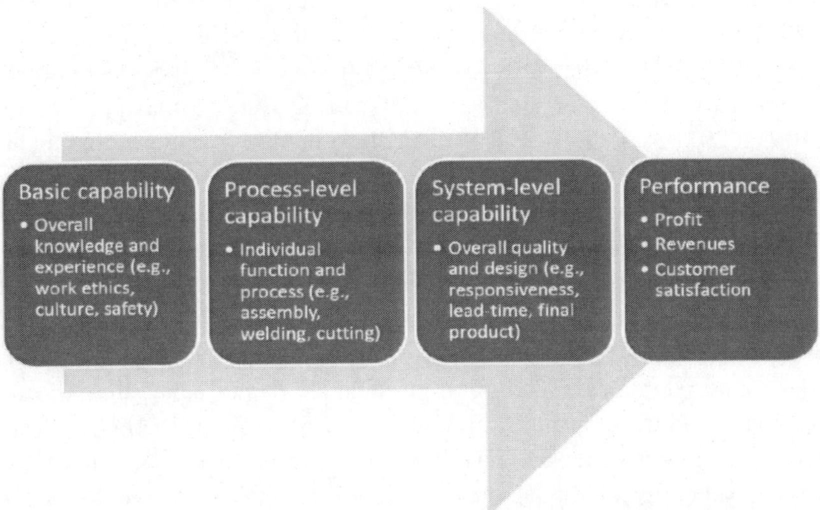

Adapted from Kim, Bowon and Chulsoon Park. (2013). "Firms' Integrating Efforts to Mitigate the Tradeoff Between Controllability and Flexibility." International Journal of Production Research. 51(4): 1258-1278.

Figure 5.3. Chain of Capability.

In addition to the tradeoff between flexibility and controllability, additive manufacturing can also directly impact a firm's chain of capability, including the basic, process-level, and system-level capabilities. At the basic level, additive manufacturing requires new knowledge, approaches, and designs. These new knowledge areas can be costly and difficult to acquire. At the process-level, a firm that adopts additive manufacturing is abandoning many of its current individual functions to adopt a radically new production method. Former functions might have required significant investment in order to fully develop. Many firms may be apprehensive in abandoning these capabilities for a new process, which itself may require significant investment to fully develop. Finally, additive manufacturing can impact the system-level capability, as it is not only a process that affects the production of individual parts, but also the assembly of the parts. All of these changes can make it costly and risky for a business to adopt additive manufacturing technologies and can result in reducing the rate at which this technology is adopted.

5.2. Adoption of Additive Manufacturing

Globally, 6494 industrial additive manufacturing systems were delivered in 2011 with a cumulative total of 49 035 systems being delivered between 1988 and 2011.[54] Of these, 18 780 were deployed in the U.S. The growth in the cumulative number of additive manufacturing systems in the U.S. between 2010 and 2011 was 15.3%.[55] It is difficult to predict the impact that additive manufacturing will have on future products. Currently, many believe that it may result in significant changes in how products are manufactured; however, there are often predictions from the past that have not come to fruition. Therefore, it is advantageous to attempt to better understand the potential future of additive manufacturing. Data from Wohlers provides some limited ability to examine past adoptions of additive manufacturing to conjecture about future adoptions.[56] The status of some of the variables that affect the adoption of additive manufacturing technologies can be observed through existing articles and texts; however, many issues cannot be substantiated without gathering additional data. Surveys can often be used to assess a producer or user's opinion of a new technology, but this is often a resource intensive process. Thomas (2013) uses domestic unit sales to estimate future adoptions of additive manufacturing.[57] Using the number of domestic unit sales[58], the growth in sales can be fitted using least squares criterion to an exponential curve that represents the traditional logistic S-curve of technology diffusion. The most widely accepted model of technology diffusion was presented by Mansfield[59]:

$$p(t) = \frac{1}{1 + e^{\alpha - \beta t}}$$

where

$p(t)$ = the proportion of potential users who have adopted the new technology by time *t*;

α = location parameter; and

β = Shape parameter (β > 0)

In order to examine additive manufacturing, it is assumed that the proportion of potential units sold by time *t* follows a similar path as the proportion of potential users who have adopted the new technology by time *t*. In order to examine shipments in the industry, it is assumed that an additive manufacturing unit represents a fixed proportion of the total revenue; thus, revenue will grow similarly to unit sales. The proportion used was calculated from 2011 data. The parameters α and β are estimated using regression on the cumulative annual sales of additive manufacturing systems in the U.S. between 1988 and 2011. United States system sales are estimated as a proportion of global sales. This method provides some insight into the current trend in the adoption of additive manufacturing technology. Unfortunately, there is little insight into the total market saturation level for additive manufacturing; that is, there is not a good sense of what percent of the relevant manufacturing industries (shown in Table 1.1) will produce parts using additive manufacturing technologies versus conventional technologies. In order to address this issue, a modified version of Mansfield's model is adopted from Chapman[60]:

$$p(t) = \frac{\eta}{1 + e^{\alpha - \beta t}}$$

where

η = market saturation level in percent.

Because η is unknown, it is varied between 0.15% and 100% of the relevant manufacturing shipments, as seen in Table 5.1. The 0.15% is derived from Wohlers estimate that the 2011 sales revenue represents 8% market penetration, which equates to $3.1 billion in market opportunity and 0.15% market saturation.

Table 5.1. Forecasts of U.S. Additive Manufacturing Shipments by Varying Market Potential

Market Potential of Relevant Manufacturing (percent of shipments)	Market Potential, Shipments ($billions 2011)	Approximate Year 100% of Market Potential Reached	Approximate Year 50% of Market Potential Reached	Approximate Year $100 Billion in Shipments is Reached	Approximate Year $50 Billion in Shipments is Reached	R^2
100.00	$2 058.9	2069	2042	2031	2028	0.948
75.00	$1 544.2	2068	2041	2031	2028	0.948
50.00	$1 029.5	2067	2039	2031	2029	0.948
45.00	$926.5	2066	2039	2031	2029	0.948
40.00	$823.6	2066	2038	2031	2029	0.948
35.00	$720.6	2065	2038	2031	2029	0.948
30.00	$617.7	2065	2037	2031	2029	0.948
25.00	$514.7	2064	2037	2032	2029	0.948
20.00	$411.8	2063	2036	2032	2029	0.948
15.00	$308.8	2062	2035	2032	2029	0.948
10.00	$205.9	2061	2033	2033	2029	0.948
5.00	$102.9	2058	2031	2044	2031	0.948
1.00	$20.6	2052	2025	-	-	0.949
0.50	$10.3	2050	2023	-	-	0.949
0.15	$3.1	2045	2018	-	-	0.950

Thomas, Douglas. 2013. Economics of the U.S. Additive Manufacturing Industry. NIST Special Publication 1163. Gaithersburg, MD: U.S. Dept. of Commerce, National Institute of Standards and Technology.

At this level, additive manufacturing is forecasted to reach 50% market potential in 2018 and 100% in 2045, as seen in the table. A more likely scenario seems to be that additive manufacturing would have between 5% and 35% market saturation. At these levels, additive manufacturing would reach 50% of market potential between 2031 and 2038 while reaching 100% between 2058 and 2065, as seen in Table 5.1. The industry would reach $50 billion between 2029 and 2031 while reaching $100 billion between 2031 and 2044.

As illustrated in Figure 5.4, it is likely that additive manufacturing is at the far left tail of the diffusion curve, making it difficult to forecast the future trends; thus, some caution should be used when interpreting this forecast. The figure illustrates the diffusion at each market saturation level presented in

Table 5.1 with the exception of the 0.50% and 0.15% levels, as they are too small to be included in this graph.

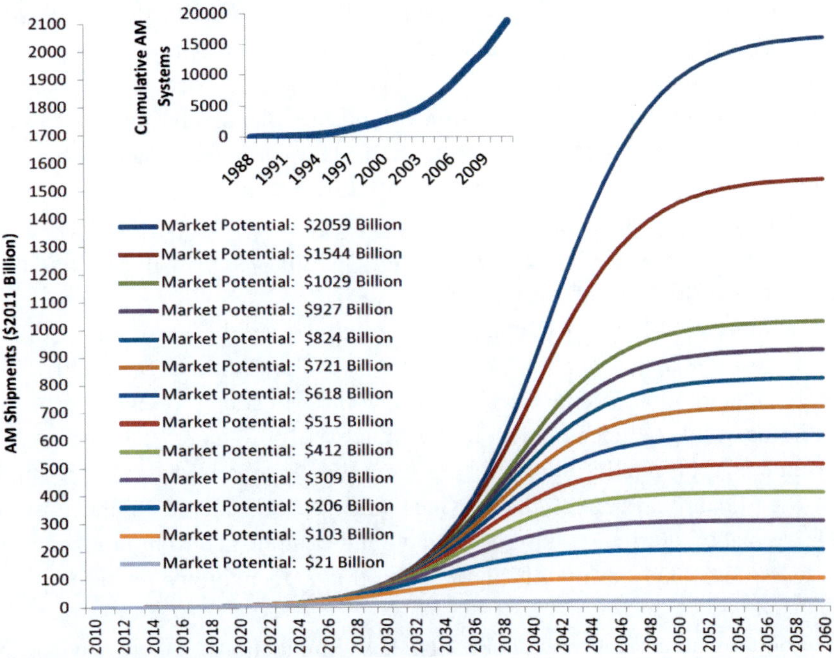

Thomas, Douglas. 2013. Economics of the U.S. Additive Manufacturing Industry. NIST Special Publication 1163. Gaithersburg, MD: U.S. Dept. of Commerce, National Institute of Standards and Technology.

Figure 5.4. Forecasts of U.S. Additive Manufacturing Shipments, by Varying Market Saturation Levels.

SUMMARY

Current research on additive manufacturing costs reveals that this technology is cost effective for manufacturing small batches with continued centralized manufacturing; however, with increased automation distributed production may be cost effective. Due to the complexities of measuring additive manufacturing costs, current studies are limited in their scope. Many of the current studies examine the production of single parts and those that examine assemblies do not examine supply chain effects such as inventory and transportation costs along with decreased risk to supply disruption. Currently, research also reveals that material costs constitute a major proportion of the cost of a product produced using additive manufacturing. Technologies can often be complementary, where two technologies are adopted alongside each other and the benefits are greater than if they were adopted individually. Increasing adoption of additive manufacturing may lead to a reduction in raw material cost through economies of scale. The reduced cost in raw material might then propagate further adoption of additive manufacturing. There may also be economies of scale in raw material costs if particular materials become more common rather than a plethora of different materials. The additive manufacturing system is also a significant cost factor; however, this cost has continually decreased. Between 2001 and 2011 the average price decreased 51% after adjusting for inflation.[61]

A number of factors complicate minimizing the cost of additive manufacturing, including build orientation, envelope utilization, build time, energy consumption, product design, and labor. The simple orientation of the part in the build chamber can result in as much as 160% increase in the energy consumed. Additionally, fully utilizing the build chamber reduces the per-unit cost significantly. Each of these issues must be considered in the cost of additive manufacturing, making it difficult and complicated to minimize costs. These issues, likely, slow the adoption of this technology, as it requires advanced knowledge.

Additive manufacturing not only has implications for the costs of production, but also the utilization of the final product. This technology allows for the manufacture of products that might not have been possible using traditional methods. These products may have new abilities, extended useful life, or reduce the time, labor, or natural resources needed to use these products. For example, automobiles might be made lighter, reducing fuel costs or combustion engines might be designed to reduce cooling needs. For this reason, there is a need to track the land (i.e., natural resources), labor, and time expended on production, utilization, and disposal along with the utility gained from new designs. The difficulty in measuring these items, likely, slows the adoption of additive manufacturing.

BIBLIOGRAPHY WITH ABSTRACTS

Allen, Jeff. 2006. "An Investigation into the Comparative Costs of Additive Manufacture vs. Machine from Solid for Aero Engine Parts." In Cost Effective Manufacture via Net- Shape Processing, 17-1 – 17-10. Meeting

Proceedings RTO-MP-AVT-139. Paper 17. DTIC Document. <http://www.rto.nato.int/abstracts.asp>

An overview of the relative economics of producing a near net shape by Additive Manufacturing (AM) processes compared with traditional machine from solid processes (MFS) is provided.

A relationship is developed to estimate the specific cost of AM material required to achieve a (typical) 30% saving over conventional MFS techniques. The use of AM techniques are shown to be advantageous for parts which have a high buy:fly ratio, have a complex shape, have a high cost of raw material used for machining from solid, have slow machining rates and are difficult and expensive to machine.

The specific cost of material deposited by additive manufacturing systems required to give a 30% saving over conventional Machine from solid techniques is estimated for a typical aerospace alloy over a range of buy:fly ratios.

The specific costs of a typical aerospace alloy deposited by present and future additive manufacturing systems are estimated and compared with the required specific costs estimated above.

It is concluded that additive manufacture is commercially viable using present additive manufacturing systems for components with a buy:fly ratio of about 12:1. For projected future additive manufacturing systems economic production of components with a buy:fly ratio of about 3 should be feasible.

Alexander, Paul, Seth Allen, and Debasish Dutta. 1998. "Part Orientation and Build Cost Determination in Layered Manufacturing." *Computer-Aided Design* 30 (5): 343–56. doi:10.1016/S0010-4485(97)00083-3.

As more choices of materials and build processes become available in layered manufacturing (LM), it is increasingly important to identify fundamental problems that underlie the entire field. Determination of best build orientation and minimizing build cost of a part are two such issues that must be considered in any LM process. By decoupling the solution to these problems from a specific LM technology, not only can the solution be applied to a variety of processes, but more realistic cost comparisons of parts built on different machines become possible.

Allen, Jeff. 2006. "An Investigation into the Comparative Costs of Additive Manufacture vs. Machine from Solid for Aero Engine Parts." DTIC Document.

An overview of the relative economics of producing a near net shape by Additive Manufacturing (AM) processes compared with traditional machine from solid processes (MFS) is provided.

A relationship is developed to estimate the specific cost of AM material required to achieve a (typical) 30% saving over conventional MFS techniques. The use of AM techniques are shown to be advantageous for parts which have a high buy:fly ratio, have a complex shape, have a high cost of raw material used for machining from solid, have slow machining rates and are difficult and expensive to machine.

The specific cost of material deposited by additive manufacturing systems required to give a 30% saving over conventional Machine from solid techniques is estimated for a typical aerospace alloy over a range of buy:fly ratios.

The specific costs of a typical aerospace alloy deposited by present and future additive manufacturing systems are estimated and compared with the required specific costs estimated above.

It is concluded that additive manufacture is commercially viable using present additive manufacturing systems for components with a buy:fly ratio of about 12:1. For projected future additive manufacturing systems economic production of components with a buy:fly ratio of about 3 should be feasible.

ATKINS Project. 2007. Manufacturing a Low Carbon Footprint: Zero Emission Enterprise Feasibility Study. Project No: N0012J. Loughborough University.

Abstract unavailable.

Atzeni, Eleonora, Luca Iuliano, Paolo Minetola, and Alessandro Salmi. 2010. "Redesign and Cost Estimation of Rapid Manufactured Plastic Parts." *Rapid Prototyping Journal* 16 (5): 308–17.

Purpose – The purpose of this paper is to highlight how rapid manufacturing (RM) of plastic parts combined with part redesign could have positive repercussion on cost saving.

Design/methodology/approach – Comparison between two different technologies for plastic part production, the traditional injection molding (IM) and the emergent RM, is done with consideration of both the geometric possibilities of RM and the economic aspect. From an extended literature review, the redesign guidelines and the cost model are identified and then applied to a component selected for its shape complexity. It is an assembly that was redesigned for RM purpose, in order to take advantage of additive manufacturing potentialities. The geometric and economic differences between IM and RM are discussed.

Findings – This research evidences that currently in Western Europe RM combined with redesign can be economically convenient and competitive to IM for medium volume production of plastic parts. Consequently, this is a great opportunity to keep the production in Europe instead of moving it overseas.

Research limitations/implications – As regards manufacturing costs, results presented in this study are mainly based on cost estimation provided by Italian companies and it is assumed that the plant is located in Western Europe.

Practical implications – The research assesses the feasibility of making functional and operational plastic parts without the use of traditional manufacturing processes by redesign for RM.

Originality/value – Two different kinds of research papers comparing RM and IM exist in literature: on the one hand, the two techniques are evaluated from the economical point of view, on the other, the part redesign is analyzed. No paper considers the interrelation between redesign and cost estimation. In this work, these aspects are combined to point out that a remarkable cost reduction is obtained when the component shape is modified to exploit RM advantages.

Atzeni, Eleonora, and Alessandro Salmi. 2012. "Economics of Additive Manufacturing for End-Usable Metal Parts." *The International Journal of Advanced Manufacturing Technology* 62 (9-12): 1147–55.

Additive manufacturing (AM) of metal parts combined with part redesign has a positive repercussion on cost saving. In fact, a remarkable cost reduction can be obtained if the component shape is modified to exploit AM potentialities. This paper deals with the evaluation of the production volume for which AM techniques result competitive with respect to conventional processes for the production of end-usable metal parts. For this purpose, a comparison between two different technologies for metal part fabrication, the traditional high-pressure die-casting and the direct metal laser sintering additive technique, is done with consideration of both the geometric possibilities of AM and the economic point of view. A design for additive manufacturing approach is adopted. Costs models of both processes are identified and then applied to an aeronautical component selected as case study. This research evidences that currently additive techniques can be economically convenient and competitive to traditional processes for small to medium batch production of metal parts.

Atzeni, Eleonora, Luca Iuliano, and Allessandro Salmi. 2011. "On the Competitiveness of Additive Manufacturing for the Production of Metal

Parts." 9th International Conference on Advanced Manufacturing Systems and Technology.

Additive Manufacturing (AM) of metal parts combined with part redesign has positive repercussion on cost saving. In fact a remarkable cost reduction can be obtained if the component shape is modified to exploit AM potentialities. This paper deals with the evaluation of the production volume for which AM techniques result competitive with respect to conventional processes. For this purpose a comparison between two different technologies for metal part production, the traditional high pressure die casting (HPDC) and the innovative AM, is done with consideration of both the geometric possibilities of AM and the economic point of view. Redesign guidelines and costs models are identified and then applied to an aeronautical component selected as case study. This research evidences that currently additive techniques can be economically convenient and competitive to traditional processes for low volume production of metal parts.

Baldinger, M., and A. Duchi. 2013. "Price Benchmark of Laser Sintering Service Providers." In High Value Manufacturing: Advanced Research in Virtual and Rapid Prototyping: Proceedings of the 6th International Conference on Advanced Research in Virtual and Rapid Prototyping, Leiria, Portugal, 1-5 October, 2013, 37. Leiria, Portugal: CRC Press.

Additive manufacturing is not only use for rapid prototyping in product development but increasingly for rapid manufacturing – meaning for production of final parts. Besides limitations around materials, quality and standards, cost is one of the major barriers to more widespread adoption. Due to the high investment for rapid manufacturing equipment and lack of knowledge, many companies choose to buy instead of make additively manufactured parts. Despite the importance of cost, there is limited insight into the price structure of additive manufacturing service providers. This study aims to narrow this gap through global price benchmark of labor sintering service providers.

Bartolo, Paulo Jorge da Silva, Mateus Artur Jorge, Fernando da Conceicao Batista, Henrique Amorim Almeida, Joao Manuel Matias, Joel Correia Vasco, Jorge Brites Gaspar, et al., eds. 2007. *Virtual and Rapid Manufacturing: Advanced Research in Virtual and Rapid Prototyping*. CRC Press.

Collection of 120 peer-reviewed papers that were presented at the 3rd International Conference on Advanced Research in Virtual and

Rapid Prototyping, held in Leiria, Portugal in September 2007. Essential reading for all those working on V&RP, focused on inducing increased collaboration between industry and academia. In addition to keynotes dealing with cutting-edge manufacturing engineering issues, contributions deal with topical research virtual and rapid prototyping (V&RP), such as: 1. biomanufacturing, 2. CAD and 3D data acquistion technologies, 3. materials, 4. Rapid tooling and manufacturing, 6. advanced rapid prototyping technologies and nanofabrication, 7. virtual environments, 8. collaborative design and engineering and 9. various applications.

Baumers, Martin. 2012. "Economic Aspects of Additive Manufacturing: Benefits, Costs and Energy Consumption".

Additive Manufacturing (AM) refers to the use of a group of technologies capable of combining material layer-by-layer to manufacture geometrically complex products in a single digitally controlled process step, entirely without moulds, dies or other tooling. AM is a parallel manufacturing approach, allowing the contemporaneous production of multiple, potentially unrelated, components or products. This thesis contributes to the understanding of the economic aspects of additive technology usage through an analysis of the effect of AM s parallel nature on economic and environmental performance measurement. Further, this work assesses AM s ability to efficiently create complex components or products. To do so, this thesis applies a methodology for the quantitative analysis of the shape complexity of AM output. Moreover, this thesis develops and applies a methodology for the combined estimation of build time, process energy flows and financial costs. A key challenge met by this estimation technique is that results are derived on the basis of technically efficient AM operation. Results indicate that, at least for the technology variant Electron Beam Melting, shape complexity may be realised at zero marginal energy consumption and cost. Further, the combined estimator of build time, energy consumption and cost suggests that AM process efficiency is independent of production volume. Rather, this thesis argues that the key to efficient AM operation lies in the user s ability to exhaust the available build space.

Baumers, M., C. Tuck, R. Hague, I. Ashcroft, and R. Wildman. 2010. "A Comparative Study of Metallic Additive Manufacturing Power Consumption." In *21st Annual International Solid Freeform Fabrication Symposium–An Additive Manufacturing Conference, Austin/TX/USA, 9th–11th August*. Austin, TX.

Efficient resource utilisation is seen as one of the advantages of Additive Manufacturing (AM). This paper presents a comparative assessment of electricity consumption of two major metallic AM processes, selective laser melting and electron beam melting. The experiments performed for this study are based on the production of a common power monitoring geometry. Due to the technology's parallel nature, the degree of build volume utilization will affect any power consumption metric.

Therefore, this work explores energy consumption on the basis of whole builds - while compensating for discrepancies in packing efficiency. This provides insight not only into absolute levels of power consumption but also on comparative process efficiency.

Baumers, M., C. Tuck, R. Wildman, I. Ashcroft, and R. Hague. 2011. "Energy Inputs to Additive Manufacturing: Does Capacity Utilization Matter?" In *22nd Annual International Solid Freeform Fabrication Symposium–An Additive Manufacturing Conference, Austin/TX/USA, 8th–10th August.*

The available additive manufacturing (AM) platforms differ in terms of their operating principle, but also with respect to energy input usage. This study presents an overview of electricity consumption across several major AM technology variants, reporting specific energy consumption during the production of dedicated test parts (ranging from 61 to 4849 MJ per kg deposited). Applying a consistent methodology, energy consumption during single part builds is compared to the energy requirements of full build experiments with multiple parts (up to 240 units). It is shown empirically that the effect of capacity utilization on energy efficiency varies strongly across different platforms.

Baumers, M., C. Tuck, R. Wildman, I. Ashcroft, E. Rosamond, and R. Hague. 2012. "Combined Buildtime, Energy Consumption and Cost Estimation for Direct Metal Laser Sintering." In *23rd Annual International Solid Freeform Fabrication Symposium–An Additive Manufacturing Conference, Austin/TX/USA, 6th–8th August.*

As a single-step process, Additive Manufacturing (AM) affords full measurability with respect to process energy inputs and production cost. However, the parallel character of AM (allowing the contemporaneous production of multiple parts) poses a number of problems for the estimation of resource consumption. A novel combined estimator of build-time, energy consumption and production cost is presented for the EOSINT M270 Direct Metal Laser Sintering system. It is demonstrated that the quantity and variety of parts demanded and the resulting ability to utilize the available machine capacity impact process efficiency, both in energy and in financial terms.

Baumers, Martin, Chris Tuck, Ricky Wildman, Ian Ashcroft, Emma Rosamond, and Richard Hague. 2013. "Transparency Built-in Energy Consumption and Cost Estimation for Additive Manufacturing." *Journal of Industrial Ecology* 17 (3): 418–31. doi:10.1111/j.1530-9290.2012.00512.x.

The supply chains found in modern manufacturing are often complex and long. The resulting opacity poses a significant barrier to the measurement and minimization of energy consumption and therefore to the implementation of sustainable manufacturing. The current article investigates whether the adoption of additive manufacturing (AM) technology can be used to reach transparency in terms of energy and financial inputs to manufacturing operations. AM refers to the use of a group of electricity- driven technologies capable of combining materials to manufacture geometrically complex products in a single digitally controlled process step, entirely without molds, dies, or other tooling. The single-step nature affords full measurability with respect to process energy inputs and production costs. However, the parallel character of AM (allowing the contemporaneous production of multiple parts) poses previously unconsidered problems in the estimation of manufacturing resource consumption. This research discusses the implementation of a tool for the estimation of process energy flows and costs occurring in the AM technology variant direct metal laser sintering. It is demonstrated that accurate predictions can be made for the production of a basket of sample parts. Further, it is shown that, unlike conventional processes, the quantity and variety of parts demanded and the resulting ability to fully utilize the available machine capacity have an impact on process efficiency. It is also demonstrated that cost minimization in additive manufacturing may lead to the minimization of process energy consumption, thereby motivating sustainability improvements.

Behdani, Behzad, Zofia Lukszo, Arief Adhitya, and Rajagopalan Srinivasan. 2009. "Agent-Based Modeling to Support Operations Management in a Multi-Plant Enterprise." In *Proceedings of the 2009 IEEE International Conference on Networking, Sensing and Control, Okayama, Japan, March 26-29, 2009*, 323–28. Okayama, Japan: IEEE.

A global industrial enterprise is a complex network of different distributed production plants producing, handling, and distributing specific products. Agent-based modeling is a proven approach for modeling complex networks of intelligent and distributed actors. In this paper we will demonstrate how an agent-based model can be used to evaluate the dynamic behavior of a global enterprise, considering both the system-level performance as well as the components' behavior. Such quantitative model can be very useful for predicting the effects of local and operational activities on plant performance and improving the tactical and strategic decisionmaking at the enterprise level.

Byun, Hong S., and Kwan H. Lee. 2006. "Determination of Optimal Build Direction in Rapid Prototyping with Variable Slicing." *The International Journal of Advanced Manufacturing Technology* 28 (3-4): 307–13. doi:10.1007/s00170-004-2355-5.

Several important factors must be taken into consideration in order to maximize the efficiency of rapid prototyping (RP) processes. The ability to select the optimal orientation of the build direction is one of the most critical factors in using RP processes, since it affects part quality, build time, and part cost. This study aims to determine the optimal build-up direction when a part is built with the variable layer thickness for different RP systems. The average weighted surface roughness (AWSR) that is generated from the stair stepping effect, the build time, and the part cost using the variable layer thickness are all considered in the process. Using the multi-attribute decision-making method, the best orientation is determined among the orientation candidates chosen from the convex hull of a model. The validity of the algorithm is illustrated by an example. The algorithm can help RP users select the best build-up direction of the part and create an efficient process planning.

Campbell, I., J. Combrinck, D. De Beer, and L. Barnard. 2008. "Stereolithography Build Time Estimation Based on Volumetric Calculations. Rapid Prototyping Journal. 14(5): 271-279.

Purpose – Not all the inventors and designers have access to computer-aided design (CAD) software to transform their design or invention into a 3D solid model. Therefore, they cannot submit an STL file to a rapid prototyping (RP) service bureau for a quotation but perhaps only a 2D sketch or drawing. This paper proposes an alternative approach to build time estimation that will enable cost quotations to be issued before 3D CAD has been used.

Design/methodology/approach – The study presents a method of calculating build time estimations within a target error limit of 10 per cent of the actual build time of a prototype. This is achieved by using basic volumetric shapes, such as cylinders and cones, added together to

represent the model in the 2D sketch. By using this information the build time of the product is then calculated with the aid of models created in a mathematical solving software package.

Findings – The development of the build time estimator and its application to several build platforms are described together with an analysis of its performance in comparison with the benchmark software. The estimator was found to meet its target 10 per cent error limit in 80 per cent of the stereolithography builds that were analysed.

Research limitations/implications – The estimator method was not able to handle multi- component complex parts builds in a timely manner. There is a trade-off between accuracy and processing time.

Practical implications – The output from the estimator can be fed directly into cost quotations to be sent to RP bureau customers at a very early stage in the design process.

Originality/value – Unlike all the other build estimators that were encountered, this method works directly from a 2D sketch or drawing rather than a 3D CAD file.

Chapman, Robert. "Benefits and Costs of Research: A Case Study of Construction Systems Integration and Automation Technologies in Commercial Buildings." NISTIR 6763. December 2001. National Institute of Standards and Technology.

This report focuses on a critical analysis of the economic impacts of past, ongoing, and planned research of the NIST Building and Fire Research Laboratory (BFRL) construction systems integration and automation technologies (CONSIAT) program. The CONSIAT program is an interdisciplinary research effort within BFRL - in collaboration with the Construction Industry Institute, the private sector, other federal agencies, and other laboratories within NIST - to develop key enabling technologies, standard communication protocols, and advanced measurement technologies needed to deliver fully- integrated and automated project process (FIAPP) products and services to the construction industry. The results of this analysis demonstrate that the use of FIAPP products and services will generate substantial cost savings to the owners and managers of commercial buildings and a contractors engaged in the construction of those buildings. The present value of savings nationwide expected from the use of FIAPP products and services is nearly $1.4 billion (measured in 1997 dollars). Furthermore, because of BFRL's involvement, FIAPP products and services are expected to be commercially available in 2005. If BFRL had not participated in the development of FIAPP products and services, the commercial introduction of FIAPP products and services is expected to be delayed until 2009. Consequently, potential cost savings accruing to *commercial building owners and managers and to contractors over the period 2005 through 2008 would have been foregone. The present value of these cost savings is approximately $120 million. These cost savings measure the value of BFRL's contribution for its CONSIAT-related investment costs of approximately $29.1 million. Stated in present value terms, every public dollar invested in BFRL's CONSIAT-related research, development, and deployment effort is expected to generate $4.13 in cost savings to the public.*

Chen, Calvin C., and Paul A. Sullivan. 1996. "Predicting Total Build-Time and the Resultant Cure Depth of the 3D Stereolithography Process." *Rapid Prototyping Journal* 2 (4): 27–40. doi:10.1108/13552549610153389.

Accurate build-time prediction for making stereolithography parts not only benefits the service industry with information necessary for correct pricing and effective job scheduling, it also provides researchers with valuable information for various build parameter studies. Instead of the conventional methods of predicting build time based on the part's volume and surface, the present predictor uses the detailed scan and recoat information from the actual build files by incorporating the algorithms derived from a detailed study of the laser scan mechanism of the stereolithography machine. Finds that the scan velocity generated from the stereolithography machine depends primarily on the system's laser power, beam diameter, materials properties and the user's specification of cure depth. Proves that this velocity is independent of the direction the laser travels, and does not depend on the total number of segments of the scan path. In addition, the time required for the laser to jump from one spot to another without scan is linearly proportional to the total jump distance, and can be calculated by a proposed constant velocity. Most profoundly, the present investigation concludes that the machine uses a velocity factor which is only 68.5 per cent of the theoretical calculation. This much slower velocity results in an undesired amount of additional cure and proves to be the main cause of the Z dimensional inaccuracy. The present build-time predictor was developed by taking into account all the factors stated above, and its accuracy was further verified by comparing the actual build-time observed for many jobs over a six month period.

Choi, S. H, and S Samavedam. 2002. "Modelling and Optimisation of Rapid Prototyping." *Computers in Industry* 47 (1): 39–53. doi:10.1016/S0166-3615(01)00140-3.

This paper proposes a Virtual Reality (VR) system for modelling and optimisation of Rapid Prototyping (RP) processes. The system aims to reduce the manufacturing risks of prototypes early in a product development cycle, and hence, reduces the number of costly design-build-test cycles. It involves modelling and simulation of RP in a virtual system, which facilitates visualisation and testing the effects of process parameters on the part quality. Modelling of RP is based on quantifying the measures of part quality, which includes accuracy, build-time and efficiency with orientation, layer thickness and hatch distance. A mathematical model has been developed to estimate the build-time of the Selective Laser Sintering (SLS) process. The model incorporates various process parameters like layer thickness, hatch space, bed temperatures, laser power and sinter factor, etc. It has been integrated with the virtual simulation system to provide a test-bed to optimise the process parameters.

Di Angelo, Luca, and Paolo Di Stefano. 2011. "A Neural Network-Based Build Time Estimator for Layer Manufactured Objects." *International Journal of Advanced Manufacturing Technology* 57 (1-4): 215–24. doi:10.1007/s00170-011-3284-8.

A correct prediction of build time is essential to calculate the accurate cost of a layer manufactured object. The methods presented in literature are of two types: detailed-analysis- and parametric-based approaches. The former require that a lot of data, related to the kinematic and dynamic performance of the machine, should be known. Parametric models, on the other hand, are of general use and relatively simple to implement; however, the parametric methods presented in literature only provide a few of the components of the total build time. Therefore, their performances are not properly suited in any case. In order to overcome these limitations, this paper proposes a parametric approach which uses a more complete set of build-time driving factors. Furthermore, considering the complexity of the parametric build time function, an artificial neural network is used so as to improve the method flexibility. The analysis of the test cases shows that the proposed approach provides a quite accurate estimation of build time even in critical cases and when supports are required.

Diegel, Olaf, Sarat Singamneni, Stephen Reay, and Andrew Withell. 2010. "Tools for Sustainable Product Design: Additive Manufacturing." *Journal of Sustainable Development* 3 (3).

The advent of additive manufacturing technologies presents a number of opportunities that have the potential to greatly benefit designers, and contribute to the sustainability of products. Additive manufacturing technologies have removed many of the manufacturing restrictions that may previously have compromised a designer's ability to make the product they imagined. Products can also be extensively customized to the user thus, once again, potentially increasing their desirability, pleasure and attachment and therefore their longevity. As additive manufacturing technologies evolve, and more new materials become available, and multiple material technologies are further developed, the field of product design has the potential to greatly change. This paper examines how aspects of additive manufacturing, from a sustainable design perspective, could become a useful tool in the arsenal to bring about the sustainable design of consumer products.

Dietrich, David M., and Elizabeth Cudney. 2011. "Impact of Integrative Design on Additive Manufacturing Quality." *International Journal of Rapid Manufacturing* 2 (3):121–31.

To move additive manufacturing (AM) into a realm of credible manufacturing, quality evaluation techniques must be established to highlight the potential gains of AM technologies in the field of production quality in terms of dimensional control. This research aims to express the relationship among AM-enabled integrative design and quality evaluation techniques. The methodology proposed is backed by a comprehensive literature review that covers AM dimensional quality and conventional quality assessment techniques for production. The research proposes modelling the positive impact of integrating design using Taguchi's quality loss function (QLF) and tolerance stack-up models. In addition, the research provides a straightforward way to evaluate AM-enabled integrated designs that promotes the proliferation of AM technology as a sustainable and credible manufacturing method. A case study is presented that describes how to apply Taguchi?s QLF to AM integrated designs.

Direct Manufacturing Research Center. "Project CoA2MPLy: Costing Analysis for Additive Manufacturing (AM) during Product Lifecycle."

Abstract unavailable.

Doubrovski, Zjenja, Jouke C. Verlinden, and Jo MP Geraedts. 2011. "Optimal Design for Additive Manufacturing: Opportunities and Challenges." In Proceedings of the ASME 2011 International Design Engineering Technical Conferences & Computers and Information in Engineering Conference IDETC/CIE 2011 August 29-31, 2011, Washington, DC,

USA, 635–46. Washington, DC, USA: American Society of Mechanical Engineers.

Additive Manufacturing (AM) represents a maturing collection of production technologies also known as rapid prototyping, rapid manufacturing and three-dimensional printing. One of the most promising aspects of AM is the possibility to create highly complex geometries. Despite a growing body of knowledge concerning the technological challenges, there is a lack of methods that allow designers to effectively deal with the new possibilities.

This article presents a literature survey on the impact that AM can have on design. The survey was focused on the new opportunities of fabrication processes, the relationship between structure and performance, and optimization approaches. We applied Olsen's three-link chain model to relate product structure with performance, linked by strength, stiffness, compliance, dynamic, thermal, and visual properties. We also use this model to base our proposed Design for Additive Manufacturing (DfAM) method.

The findings show that there is a growing body of knowledge in the field of design for AM (DfAM), yet only considers a subset of properties. Furthermore, the knowledge on materials, computational optimization, computer aided design, and behavioral simulation embody separated domains and related software support. This is in contrast with design engineering, which requires a holistic approach to conceptualize new products.

Economist. Feb 18th 2010 "Printing Body Parts: Making a Bit of Me." <http://www.economist.com/node/15543683>

Abstract unavailable

Fogliatto, Flavio S, and Giovani J. C Da Silveira. 2011. *Mass Customization Engineering and Managing Global Operations*. London: Springer.

The analysis and implementation of mass customization (MC) systems has received growing consideration by researchers and practitioners since the late 1980s. In this paper we update the literature review on MC presented in a previous paper (Da Silveira, G., Borenstein, D., Fogliatto, F.S., 2001. Mass customization: literature review and research directions. International Journal of Production Economics, 72 (1), 1-13), and identify research gaps to be investigated in the future. Major areas of research in MC, and journals in which works have been published are explored through summary statistics. The result is a concise compendium of the relevant literature produced on the topic in the past decade.

Fogliatto, Flavio S., Giovani J. C. da Silveira, and Denis Borenstein. 2012. "The Mass Customization Decade: An Updated Review of the Literature." *International Journal of Production Economics* 138 (1): 14–25. doi:10.1016/j.ijpe.2012.03.002.

Mass customization (MC) has been hailed as a successful operations strategy across manufacturing and service industries for the past three decades. However, the wider implications of using MC approaches in the broader industrial and economic environment are not yet clearly understood. Mass Customization: Engineering and Managing Global Operations presents emerging research on the role of MC and personalization in today's international operations context. The chapters cover MC in the context of global industrial economics and operations. Moreover, the book discusses MC topics that are relevant.

Giannatsis, J, V Dedoussis, and L Laios. 2001. "A Study of the Build-Time Estimation Problem for Stereolithography Systems." *Robotics and Computer-Integrated Manufacturing* 17 (4): 295–304. doi:10.1016/S0736-5845(01)00007-2.

In this paper the problem of build-time estimation for Stereolithography systems is examined. Experimental results from various case studies indicate that the accuracy of estimation greatly depends on the type of part geometry representation processed and the uncontrolled laser power fluctuations. It is shown that estimation based on sliced (CLI) representation can be extremely accurate, assuming that the average laser power during fabrication can be predicted. On the other hand, estimations based on tessellated (STL) representation, although not so accurate, satisfy the accuracy requirements imposed at early stages of the Stereolithography process, where no slice data are available. As part of this study, build-time itself is also analyzed and factors affecting it are identified and investigated experimentally. Results indicate that hatching time depends not only on the hatching distance and speed, as originally assumed, but also on the number of hatching vectors employed.

Gibson, Ian, David W. Rosen, and Brent Stucker. 2010. *Additive Manufacturing Technologies*. Springer.

Additive Manufacturing Technologies: Rapid Prototyping to Direct Digital Manufacturing deals with various aspects of joining materials to form parts. Additive Manufacturing (AM) is an automated technique for direct conversion of 3D CAD data into physical objects using a variety of approaches. Manufacturers have been using these technologies in order to reduce development cycle times and get their products to the market quicker, more cost effectively, and with added value due to the incorporation of customizable features. Realizing the potential of AM applications, a large number of processes have been developed allowing the use of various materials ranging from plastics to metals for product development. Authors Ian Gibson, David W. Rosen and Brent Stucker explain these issues, as well as:

-Providing a comprehensive overview of AM technologies plus descriptions of support technologies like software systems and post-processing approaches

-Discussing the wide variety of new and emerging applications like micro-scale AM, medical applications, direct write electronics and Direct Digital Manufacturing of end-use components

-Introducing systematic solutions for process selection and design for AM Additive Manufacturing Technologies: Rapid Prototyping to Direct Digital Manufacturing is the perfect book for researchers, students, practicing engineers, entrepreneurs, and manufacturing industry professionals interested in additive manufacturing.

T. A. Grimm & Associates, Inc. 2010. *3D Printer Benchmark: North American Edition*. Edgewood, KY: T. A. Grimm & Associates, Inc.

Abstract unavailable

Hasan, S., and A.E.W. Rennie. 2008. "The Application of Rapid Manufacturing Technologies in the Spare Parts Industry." In 19th Annual International Solid Freeform Fabrication Symposium–An Additive Manufacturing Conference, Austin/TX/USA, 4th–6th August. Austin, TX.

The advancement of Rapid Manufacturing (RM) has ushered the possibility of realising complex designs. This paper identifies the potential of possible applications of RM in the spare parts industry. It further underlines the need for a fully functional RM supply chain before proposing an e-business enabled business model for RM technologies.

Holmström, Jan, and Jouni Partanen. 2014. "Digital Manufacturing-Driven Transformations of Service Supply Chains for Complex Products." *Supply Chain Management: An International Journal* 19 (4): 421 – 430.

Purpose – The purpose of this paper is to explore the forms that combinations of digital manufacturing, logistics and equipment use are likely to take and how these novel combinations may affect the relationship among logistics service providers (LSPs), users and manufacturers of equipment.

Design/methodology/approach – Brian Arthur's theory of combinatorial technological evolution is applied to examine possible digital manufacturing-driven transformations. The F-18 Super Hornet is used as an illustrative example of a service supply chain for a complex product.

Findings – The introduction of digital manufacturing will likely result in hybrid solutions, combining conventional logistics, digital manufacturing and user operations. Direct benefits can be identified in the forms of life cycle extension and the increased availability of parts in challenging locations. Furthermore, there are also opportunities for both equipment manufacturers and LSPs to adopt new roles, thereby supporting the efficient and sustainable use of digital manufacturing.

Research limitations/implications – The phenomenon of digital manufacturing-driven transformations of service supply chains for complex product does not yet fully exist in the real world, and its study requires cross-disciplinary collaboration. Thus, the implication for research is to use a design science approach for early-stage explorative research on the form and function of novel combinations.

Practical implications – Digital manufacturing as a general-purpose technology gives LSPs an opportunity to consolidate demand from initial users and incrementally deploy capacity closer to new users. Reengineering the products that a manufacture currently uses is needed to increase the utilization of digital manufacturing.

Originality/value – The authors outline a typology of digital manufacturing-driven transformations and identify propositions to be explored in further research and practice.

Holmström, Jan, Jouni Partanen, Jukka Tuomi, and Manfred Walter. 2010. "Rapid Manufacturing in the Spare Parts Supply Chain: Alternative Approaches to Capacity Deployment." *Journal of Manufacturing Technology Management* 21 (6): 687–97. doi:10.1108/17410381011063996.

Purpose – The purpose of this paper is to describe and evaluate the potential approaches to introduce rapid manufacturing (RM) in the spare parts supply chain.

Design/methodology/approach – Alternative conceptual designs for deploying RM technology in the spare parts supply chain were proposed. The potential benefits are illustrated for the aircraft industry. The general feasibility was discussed based on literature.

Findings – The potential supply chain benefits in terms of simultaneously improved service and reduced inventory makes the distributed deployment of RM very interesting for spare parts supply. However, considering the trade-offs affecting deployment it is proposed that most feasible is centralized deployment by original equipment manufacturers (OEMs), or deployment close to the point of use by generalist service providers of RM.

Research limitations/implications – The limited part range that is currently possible to produce using the technology means that a RM-based service supply chain is feasible only in very particular situations.

Practical implications – OEMs should include the consideration of RM in their long-term service supply chain development.

Originality/value – The paper identifies two distinct approaches for deploying RM in the spare parts supply chain.

Hopkinson, Neil. 2006. "Production Economics of Rapid Manufacture." In *Rapid Manufacturing: An Industrial Revolution for the Digital Age*, 147–57.

Abstract unavailable

Hopkinson, Neil, and P. Dickens. 2003. "Analysis of Rapid Manufacturing—using Layer Manufacturing Processes for Production." *Proceedings of the Institution of Mechanical Engineers, Part C: Journal of Mechanical Engineering Science* 217 (1): 31–39.

Rapid prototyping (RP) technologies that have emerged over the last 15 years are all based on the principle of creating three-dimensional geometries directly from computer aided design (CAD) by stacking two-dimensional profiles on top of each other. To date most RP parts are used for prototyping or tooling purposes; however, in future the majority may be produced as end-use products. The term 'rapid manufacturing' in this context uses RP technologies as processes for the production of end-use products.

This paper reports findings from a cost analysis that was performed to compare a traditional manufacturing route (injection moulding) with layer manufacturing processes (stereolithography, fused deposition modelling and laser sintering) in terms of the unit cost for parts made in various quantities. The results show that, for some geometries, it is more economical to use layer manufacturing methods than it is to use traditional approaches for production in the thousands.

Hopkinson, Neil, Richard Hague, and Philip Dickens, eds. 2006. *Rapid Manufacturing: An Industrial Revolution for the Digital Age*. John Wiley & Sons.

Rapid Manufacturing is a new area of manufacturing developed from a family of technologies known as Rapid Prototyping. These processes have already had the effect of both improving products and reducing their development time; this in turn resulted in the development of the technology of Rapid Tooling, which implemented Rapid Prototyping techniques to improve its own processes. Rapid Manufacturing has developed as the next stage, in which the need for tooling is eliminated. It has been shown that it is economically feasible to use existing commercial Rapid Prototyping systems to manufacture series parts in quantities of up to 20,000 and customised parts in quantities of hundreds of thousands. This form of manufacturing can be incredibly cost-effective and the process is far more flexible than conventional manufacturing.

Rapid Manufacturing: An Industrial Revolution for the Digital Age addresses the academic fundamentals of Rapid Manufacturing as well as focussing on case studies and applications across a wide range of industry sectors. As a technology that allows manufacturers to create products without tools, it enables previously impossible geometries to be made. This book is abundant with images depicting the fantastic array of products that are now being commercially manufactured using these technologies.

-Includes contributions from leading researchers working at the forefront of industry.

-Features detailed illustrations throughout.

Rapid Manufacturing: An Industrial Revolution for the Digital Age is a groundbreaking text that provides excellent coverage of this fast emerging industry. It will interest manufacturing industry practitioners in research and development, product design and materials science, as well as having a theoretical appeal to researchers and post-graduate students in manufacturing engineering, product design, CAD/CAM and CIFM.

Huang, Samuel H., Peng Liu, Abhiram Mokasdar, and Liang Hou. 2013. "Additive Manufacturing and Its Societal Impact: A Literature Review." *The International Journal of Advanced Manufacturing Technology* 67 (5-8): 1191–1203. doi:10.1007/s00170-012-4558-5.

Thirty years into its development, additive manufacturing has become a mainstream manufacturing process. Additive manufacturing build up parts by adding materials one layer at a time based on a computerized 3D solid model. It does not require the use of fixtures, cutting tools, coolants, and other auxiliary resources. It allows design optimization and the producing of customized parts on-demand. Its advantages over conventional manufacturing have captivated the imagination of the public, reflected in recent mainstream publications that call additive manufacturing "the third industrial revolution." This paper reviews the societal impact of additive manufacturing from a technical perspective. Abundance of evidences were found to support the promises of additive manufacturing in the following areas: (1) customized healthcare products to improve population health and quality of life, (2) reduced environmental impact for manufacturing sustainability, and (3) simplified supply chain to increase efficiency and responsiveness in demand fulfillment. In the mean time, the review also identified the need for further research in the areas of life-cycle energy consumption evaluation and potential occupation hazard assessment for additive manufacturing.

Igoe, Tom, and Catarina Mota. 2011. "A Strategist's Guide to Digital Fabrication." *Strategy+Business*, no. 64 (Autumn): 1–10.

Rapid advances in manufacturing technology point the way toward a decentralized, more customer- centric "maker" culture. Here are the changes to consider before this innovation takes hold.

Kechagias, John, Stergios Maropoulos, and Stefanos Karagiannis. 2004. "Process Build- Time Estimator Algorithm for Laminated Object Manufacturing." *Rapid Prototyping Journal* 10 (5): 297–304. doi:10.1108/13552540410562331.

A method for estimating the build-time required by the laminated object manufacturing (LOM) process is presented in this paper. The proposed algorithm – taking into account the real process parameters and the information included in the part's STL-file – performs a minimum manipulation of the file, and calculates total volume, total surface area and flat areas involved in fine cross-hatching. A number of experiments performed verify the applicability of the algorithm in process build-time estimation. The time prediction estimates are within 7.6 per cent of the real build-times for the LOM process. It is believed that, through specific minor adjustments, the algorithm could well be employed in process build-time estimation for similar rapid prototyping processes.

Kellens, K., E. Yasa, Renaldi, W. Dewulf, JP Kruth, and J.R. Duflou. 2011. "Analyzing Product Lifecycle Costs for a Better Understanding of Cost Drivers in Additive Manufacturing." In *22nd Annual International Solid Freeform Fabrication Symposium– An Additive Manufacturing Conference, Austin/TX/USA, 8th–10th August*. Austin, TX.

Manufacturing processes, as used for discrete part manufacturing, are responsible for a substantial part of the environmental impact of products, but are still poorly documented in terms of their environmental footprint. The lack of thorough analysis of manufacturing processes has as consequence that optimization opportunities are often not recognized and that improved machine tool design in terms of ecological footprint has only been targeted for a few common processes.

Additive manufacturing processes such as Selective Laser Sintering (SLS) and Selective Laser Melting (SLM) allow near-net shape manufacturing of complex work pieces. Consequently, they inherently offer opportunities for minimum-waste and sustainable manufacturing. Nevertheless, powder production, energy consumption as well as powder losses are important and not always optimized environmental impact drivers of SLS and SLM. This paper presents the results of a data collection effort, allowing to assess the overall environmental impact of these processes using the methodology of the CO2PE! (Cooperative Effort on Process Emissions in Manufacturing) initiative.

Based on the collected LCI data, a subsequent impact assessment analysis allows indentifying the most important contributors to the environmental impact of SLS/SLM. Next to the electricity consumption, the consumption of inert gasses proves to be an important cause of environmental impact. Finally, the paper sketches the improvement potential for SLS/SLM on machine tool as well as system level.

Kellens, Karel, Wim Dewulf, Wim Deprez, Evren Yasa, and Joost Duflou. 2010. "Environmental Analysis of SLM and SLS Manufacturing Processes." In *Proceedings of LCE2010 Conference*, 423–28. Hefei, China.

Manufacturing processes, as used for discrete part manufacturing, are responsible for a substantial part of the environmental impact of products, but are still poorly documented in terms of environmental footprint. In this paper, first a short description is offered about the CO2PE! – Initiative and the methodology used to analyse manufacturing unit processes. In a second part, the energy and resource flows inventorisation and impact assessment of some sample products made by Selective Laser Melting (SLM) and Selective Laser Sintering (SLS) processes are performed.

Kellens, Karel, Wim Dewulf, Michael Overcash, Michael Z. Hauschild, and Joost R. Duflou. 2012. "Methodology for Systematic Analysis and Improvement of Manufacturing Unit Process Life-Cycle Inventory (UPLCI)—CO2PE! Initiative (cooperative Effort on Process Emissions in Manufacturing). Part 1: Methodology Description." *The International Journal of Life Cycle Assessment* 17 (1): 69–78. doi:10.1007/s11367-011-0340-4.

Purpose This report proposes a life-cycle analysis (LCA)-oriented methodology for systematic inventory analysis of the use phase of manufacturing unit processes providing unit process datasets to be used in life-cycle inventory (LCI) databases and libraries. The methodology has been developed in the framework of the CO2PE! collaborative research programme (CO2PE! 2011a) and comprises two approaches with different levels of detail, respectively referred to as the screening approach and the in-depth approach.

Methods The screening approach relies on representative, publicly available data and engineering calculations for energy use, material loss, and identification of variables for improvement, while the in-depth approach is subdivided into four modules, including a time study, a power consumption study, a consumables study and an emissions study, in which all relevant process in- and outputs are measured and analysed in detail. The screening approach provides the first insight in the unit process and results in a set of approximate LCI data, which also serve to guide the more detailed and complete in-depth approach leading to more accurate LCI data as well as the identification of potential for energy and resource efficiency improvements of the manufacturing unit process. To ensure optimal reproducibility and applicability, documentation guidelines for data and metadata are included in both approaches. Guidance on definition of functional unit and reference flow as well as on determination of system boundaries specifies the generic goal and scope definition requirements according to ISO 14040 (2006) and ISO 14044 (2006).

Results The proposed methodology aims at ensuring solid foundations for the provision of high-quality LCI data for the use phase of manufacturing unit processes. Envisaged usage encompasses the provision of high-quality data for LCA studies of products using these unit process datasets for the manufacturing processes, as well as the in-depth analysis of individual manufacturing unit processes.

Conclusions In addition, the accruing availability of data for a range of similar machines (same process, different suppliers and machine capacities) will allow the establishment of parametric emission and resource use estimation models for a more streamlined LCA of products including reliable manufacturing process data. Both approaches have already provided useful results in some initial case studies (Kellens et al. 2009; Duflou et al. (Int J Sustain Manufacturing 2:80–98, 2010); Santos et al. (J Clean Prod 19:356–364, 2011); UPLCI 2011; Kellens et al. 2011a) and the use will be illustrated by two case studies in Part 2 of this paper (Kellens et al. 2011b).

Khajavi, Siavash H., Jouni Partanen, and Jan Holmström. 2014. "Additive Manufacturing in the Spare Parts Supply Chain." *Computers in Industry* 65 (1): 50–63.

As additive manufacturing (AM) evolves to become a common method of producing final parts, further study of this computer integrated technology is necessary. The purpose of this research is to evaluate the potential impact of additive manufacturing improvements on the configuration of spare parts supply chains. This goal has been accomplished through scenario modeling of a real-life spare parts supply chain in the aeronautics industry. The spare parts supply chain of the F-18 Super Hornet fighter jet was selected as the case study because the air-cooling ducts of the environmental control system are produced using AM technology. In total, four scenarios are investigated that vary the supply chain configurations and additive manufacturing machine specifications. The reference scenario is based on the spare parts supplier's current practice and the possible future decentralization of production and likely improvements in AM technology. Total operating cost, including downtime cost, is used to compare the scenarios. We found that using current AM technology, centralized production is clearly the preferable supply chain configuration in the case example. However, distributed spare parts production becomes practical as AM machines become less capital intensive, more autonomous and offer shorter production cycles. This investigation provides guidance for the development of additive manufacturing machines and their possible deployment in spare parts supply chains. This study contributes to the emerging literature on AM deployment in supply chains with a real-world case setting and scenario model illustrating the cost trade-offs and critical requirements for technology development.

Kim, Bowon. "Supply Chain Management: A Learning Perspective." Korea Advanced Institute of Science and Technology. Coursera Lecture 1-2.

As a human being, we all consume products and/or services all the time. This morning you got up and ate your breakfast, e.g., eggs, milk, bread, fresh fruits, and the like. After the breakfast, you drove your car to work or school. At your office, you used your computer, perhaps equipped with 27" LCD monitor. During your break, you drank a cup of coffee and played with your iPhone. So on and so forth. You probably take it for granted that you can enjoy all of these products. But if you take a closer look at how each of these products can be made and eventually delivered to you, you will realize that each one of these is no short of miracle. For example, which fruit do you like? Consider fresh strawberries. In order for the strawberries to be on your breakfast table, there must be numerous functions, activities, transactions, and people involved in planting, cultivating, delivering, and consuming strawberries. Moreover, all of these functions, activities, transactions, and people are connected as an integral chain, through which physical products like strawberries themselves and virtual elements such as information and communication flow back and forth constantly. By grouping related functions or activities, we have a supply chain, comprised of four primary functions such as supplier, manufacturer, distributor, and finally consumer. A supply chain is essentially a value chain.

For the society or economy as a whole, the goal is to maximize value, i.e., to create satisfactory value without spending too much. In order to create the maximum value for the strawberry supply chain, every participant in the chain must carry out its function efficiently. In addition, all of the members must coordinate with each other effectively in order to ensure value maximization. We have to face the same issues for almost all the products and services we take for granted in our everyday life, e.g., cars, hamburgers, haircuts, surgeries, movies, banks, restaurants, and you name it!

In this course, we want to understand fundamental principles of value creation for the consumers or the market. We try to answer questions like how the product or service is made, how the value-creating activities or functions are coordinated, who should play what leadership roles in realizing all these, and so on. As our course title hints, we approach all of these issues from a learning perspective, which is dynamic in nature and emphasizes long-term capability building rather than short-term symptomatic problem solving.

Kim, Bowon and Chulsoon Park. (2013). "Firms' Integrating Efforts to Mitigate the Tradeoff Between Controllability and Flexibility." International Journal of Production Research. 51(4): 1258-1278.

We consider three manufacturing capabilities: controllability, flexibility, and integrating capability. Controllability is a firm's ability to control its process to enhance efficiency and accuracy and to better meet specifications. Flexibility is a firm's ability to cope with uncertainty and variation, both internal and external. Integrating capability is a firm's ability to integrate and coordinate diverse functions and parts of its supply chain, embodied in overall operations effectiveness and new product innovation. We put forth two hypotheses. First, there is an inherent tradeoff between controllability and flexibility. Second, a firm's integrating effort across its supply chain enables it to overcome such a tradeoff, making it possible to improve both controllability and flexibility simultaneously. Using data from 193 manufacturing companies, we test our hypotheses. It turns out that the relationship between controllability and flexibility is convex-shaped, indicating there are two distinct regions: one in which the relationship is negative and the other, positive. Further, the firms in the positive relationship region make significantly more effort to integrate, that is to say coordinate and communicate, across their supply chains, implying that as the firm strives to integrate its supply chain functions, it can mitigate the tradeoff between controllability and flexibility to a considerable extent.

Kruth, Jean-Pierre, Ben Vandenbroucke, van J. Vaerenbergh, and Peter Mercelis. 2005. "Benchmarking of Different SLS/SLM Processes as Rapid Manufacturing Techniques." In *Int. Conf. Polymers and Moulds Innovations (PMI), Gent, Belgium, April 20-23, 2005*. Gent, Belgium.

Recently, a shift of Rapid Prototyping (RP) to Rapid Manufacturing (RM) has come up because of technical improvements of Layer Manufacturing processes. Selective Laser Sintering (SLS) and Selective Laser Melting (SLM) techniques are no longer exclusively used for prototyping and the possibility to process all kind of metals yields opportunities to manufacture real functional parts, e.g., injection moulds (Rapid Tooling).

This study examines different SLS/SLM processes with regard to conditions that become very important for manufacturing, speed and reliability. A benchmark model is developed facilitating to test these conditions and to check the process limitations. This benchmark is manufactured by five SLS/SLM machines which differ in process mechanism, powder material and optimal process parameters. To find out process accuracy, a dimensional analysis is performed and the surface roughness is measured. Besides, the benchmarks are tested for their mechanical properties such as density, hardness, strength and stiffness. Finally, speed and repeatability are discussed as important factors for manufacturing.

This paper presents the state of the art in SLS/SLM and aims at understanding the limitations of different SLS/SLM processes to form a picture of the potential manufacturing applications of these processes.

Li, Fang. 2006. "Automated Cost Estimation for 3-Axis CNC Milling and Stereolithography Rapid Phototyping." http://mspace.lib.umanitoba.ca/jspui/handle/1993/8882.

Rapid prototyping (RP) is a supplementary additive manufacturing method to the traditional Computer Numerical Controlled (CNC) machining. The selection of the manufacturing method between RP and CNC machining is currently based on qualitative analysis and engineers' experience. There are situations when parts can be produced using either of the methods. In such cases, cost will be the decisive factor. However, lack of a quantitative cost estimation method to guide the selection between RP and CNC machining makes the decision process difficult. This thesis proposes an automated cost estimator for CNC machining and Rapid Prototyping. Vertical CNC milling and Stereolithography Apparatus (SLA) RP technology are selected in specific, for cost modeling and process comparison. A binary questionnaire is designed to help estimate the CNC setup cost. An SLA build time estimator is implemented based on 3D systems' SLA3500 machine. SLA post processing cost is also investigated. Based on the developed methods, a prototype software tool was created with an output to Excel chart to facilitate the selection. Five cases have been studied with the software and the predicted results are found reasonable and effective.

Lindemann, C., U. Jahnke, M. Moi, and R. Koch. 2012. "Analyzing Product Lifecycle Costs for a Better Understanding of Cost Drivers in Additive Manufacturing." In *23rd Annual International Solid Freeform Fabrication Symposium–An Additive Manufacturing Conference, Austin/TX/USA, 6th–8th August*. Austin, TX.

The costs of additive manufactured parts often seem too high in comparison to those of traditionally manufactured parts, as the information about major cost drivers, especially for additive manufactured metal parts, is weak. Therefore, a lifecycle analysis of additive manufactured parts is needed to understand and rate the cost drivers that act as the largest contributors to unit costs, and to provide a focus for future cost reduction activities for the Additive Manufacturing (AM) technology. A better understanding of the cost structure will help to compare the AM costs with the opportunity costs of the classical manufacturing technologies and will make it easier to justify the use of AM manufactured parts. This paper will present work in progress and methodology based on a sample investigated with business process analysis / simulation and activity based costing. In addition, cost drivers associated with metal AM process will be rated.

Lindemann, C., U. Jahnke, M. Moi, and R. Koch. 2013. "Impact and Influence Factors of Additive Manufacturing on Product Lifecycle Costs." In *24th Annual International Solid Freeform Fabrication Symposium–An Additive Manufacturing Conference, Austin/TX/USA*. Austin, TX.

At first sight the direct costs of Additive Manufacturing (AM) seem too high in comparison to traditional manufacturing. Considering the whole lifecycle costs of parts changes the point of view. Due to the modification of the new production process and new supply chains during a parts lifecycle, producing companies can strongly benefit from AM. Therefore, a costing model for assessing lifecycle costs with regard to specific applications and branches has been developed. The costing model represents the advantages of AM monetary. For the evaluation of this model and the influence factors, different case studies have been performed including different approaches in part redesign. Deeper research is and will be carried out with respect to the AM building rates and the comparability of various AM machines, as these facts are hardly comparable for end users. This paper will present the methodology as well as the results of the case studies conducted over the whole product lifecycle.

Luo, Yanchun, Zhiming Ji, M.C. Leu, and R. Caudill. 1999. "Environmental Performance Analysis of Solid Freedom Fabrication Processes." In *Proceedings of the 1999 IEEE International Symposium on Electronics and the Environment, 1999. ISEE -1999*, 1–6. doi:10.1109/ISEE.1999. 765837.

This paper presents a method for analyzing the environmental performance of solid freeform fabrication (SFF) processes. In this method, each process is divided into life phases. Environmental effects of every process phase are then analyzed and evaluated based on the environmental and resource management data. These effects are combined to obtain the environmental performance of the process. The analysis of the environmental performance of SFF processes considers the characteristics of SFF technology, includes material, energy consumption, processes wastes, and disposal. Case studies for three typical SFF processes: stereolithography (SL); selective laser sintering (SLS); and fused deposition modeling (FDM) are presented to illustrate this method

Munguia, Javier, Joaquim de Ciurana, and Carles Riba. 2009 "Neural-Network-Based Model for Build-Time Estimation in Selective Laser Sintering." Proceedings of the Institution of Mechanical Engineers. Part B, Journal of Engineering Manufacture. 223(8):995-1003.

Cost assessment for rapid manufacturing (RM) is highly dependent on time estimation. Total build time dictates most indirect costs for a given part, such as labour, machine, costs, and overheads. A numberof parametric and empirical time estimators exist; however, they normally account for error rates between 20 and 35 per cent which are then translated to inaccurate final cost estimations. The estimator presented herein is based on the ability of artificial neural networds (ANNs) to learn and adapt to different cases, so that the developed model is capable of providing accurate estimates regardless of machine type or model. A simulation is performed with MATLAB to compare existing approaches for cost/time estimation for selective laser sintering (SLS). Error rates observed from the model range from 2 to 15 per cent, which shows the validity and robustness of the proposed method.

Mansfield, Edwin. *Innovation, Technology and the Economy: Selected Essays of Edwin Mansfield.* Economists of the Twentieth Century Series (Brookfield, VT: 1995, E. Elgar).

This text brings together selected essays of Edwin Mansfield, who has been engaged for almost 40 years in the economics of technical change, a field of importance for analysts and decision-makers. This text presents a quantitative analysis based largely on data collection from firms and other economic units. These essays, which include some of the most frequently cited studies in the field, are concerned with the process of industrial innovation, the nature, composition and effects of industrial research and development, the relationships between technical change, economic growth and inflation, the diffusion of innovations, international technology transfer, public policy toward civilian technology, and intellectual property protection. These topics are central to many current debates among both economic theorists and policy makers.

Mehrsai, Afshin, Hamid Reza Karimi, and Klaus-Dieter Thoben. 2013. "Integration of Supply Networks for Customization with Modularity in Cloud and Make-to-Upgrade Strategy." *Systems Science & Control Engineering* 1 (1): 28–42. doi:10.1080/21642583.2013.817959.

Today, integration of supply networks (SNs) out of heterogeneous entities is quite challenging for industries. Individualized demands are getting continuously higher values in the global business and this fact forces traditional businesses for restructuring their organizations. In order to contribute to new performances in manufacturing networks, in this paper a collaborative approach is recommended out of modularity structure, cloud computing, and make-to-upgrade concept for improving flexibility as well as coordination of entities in networks. A cloud-based framework for inbound and outbound manufacturing is introduced for complying with the production of individualized products in the turbulent global market, with local decision-makings and integrated performances. Additionally, the complementary aspects of these techniques with new features of products are conceptually highlighted. The compatibility of this wide range of theoretical concepts and practical techniques is explained here. A discrete-event simulation out of an exemplary cloud-based SN is set up to define the applicability of the cloud and the recommended strategy.

Minetola, Paolo. 2012. "The Importance of a Correct Alignment in Contactless Inspection of Additive Manufactured Parts." *International Journal of Precision Engineering and Manufacturing* 13 (2): 211–18. doi:10.1007/s12541-012-0026-2.

Nowadays products having complex freeform custom-made shapes can he fabricated without any tool by means of additive manufacturing processes. Additive manufactured parts must be inspected for quality to verify, that they meet dimensional and geometrical specifications among other requirements just as any other product. Contactless inspection carried out with optical 3D scanners is preferred to traditional pointwise measurements because of the higher amount of data retrieved in short times. A key step of the contactless inspection process is the definition of the part reference frame for the alignment of scan data. This paper considers different 3-2-1 alignments and analyze their influence on the inspection results, putting in evidence that an inattentive or inaccurate definition of the part reference frame can lead to incorrect evaluations of real part deviations.

Mognol, Pascal, Denis Lepicart, and Nicolas Perry. 2006. "Rapid Prototyping: Energy and Environment in the Spotlight." *Rapid Prototyping Journal* 12 (1): 26–34. doi:10.1108/13552540610637246.

Purpose – To discuss integration of the rapid prototyping environmental aspects with the primary focus on electrical energy consumption.

Design/methodology/approach – Various manufacturing parameters have been tested on three rapid prototyping systems: Thermojet (3DS), FDM 3000 (Stratasys) and EOSINT M250 Xtended (EOS). The objective is to select sets of parameters for reduction of electrical energy consumption. For this, a part is manufactured in several orientations and positions in the chamber of these RP systems. For each test, the electrical power is noted. Finally, certain rules are proposed to minimize this electrical energy consumption during a job.

Findings – It is important to minimize the manufacturing time but there is no general rule for optimization of electrical energy consumption. Each RP system must be tested with energy consumption considerations under the spotlight.

Research limitations/implications – The work is only based on rapid prototyping processes. The objective is to take into consideration the complete life-cycle of a rapid prototyped part: manufacturing of raw material as far as reprocessing of waste.

Practical implications – Reduction of electrical energy consumption to complete a job.

Originality/value – Currently, environmental aspects are not well studied in rapid prototyping.

Morrow, W.R., H. Qi, I. Kim, J. Mazumder, and S.J. Skerlos. 2007. "Environmental Aspects of Laser-Based and Conventional Tool and Die Manufacturing." *Journal of Cleaner Production* 15 (10): 932–43. doi:10.1016/j.jclepro.2005.11.030.

Solid Freeform Fabrication (SFF) technologies such as Direct Metal Deposition (DMD) have made it possible to eliminate environmentally polluting supply chain activities in the tooling industry and to repair and remanufacture valuable tools and dies. In this article, we investigate three case studies to reveal the extent to which DMD-based manufacturing of molds and dies can currently achieve reduced environmental emissions and energy consumption relative to conventional manufacturing pathways. It is shown that DMD's greatest opportunity to reduce the environmental impact of tool and die manufacturing will come from its ability to enable remanufacturing. Laser-based remanufacturing of tooling is shown to reduce cost and environmental impact simultaneously, especially as the scale of the tool increases.

Moylan, Shawn, John Slotwinski, April Cooke, Kevin Jurrens, and M. Alkan Donmex. 2013. *Lessons Learned in Establishing the NIST Metal Additive Manufacturing Laboratory*. NIST Technical Note 1801. Gaithersburg, MD: U.S. Dept. of Commerce, National Institute of Standards and Technology.

This publication presents a summary of lessons learned by NIST staff during establishment of the NIST Metal Additive Manufacturing Laboratory and implementation of the metal additive manufacturing capability at NIST. These lessons learned resulted from the first implementation of a metal additive manufacturing system at NIST. While the NIST experiences were with a particular metal additive manufacturing system, we believe that these lessons are relevant and have common aspects for implementing other types of metal additive manufacturing systems. The intention is that this summary document will help others to implement metal additive manufacturing capabilities in their facilities. The NIST implementation spanned several months before the system was brought fully online,

including facility preparation, system installation, operator training, standard procedure development, and initial experimental use. NIST staff members have been operating the machine for research purposes since early 2011. Parts have been built using metal powders of one stainless steel and one Cobalt-Chrome alloy. These lessons learned address room requirements, safety concerns, machine operation, materials and process parameters, build design file preparation and support structures, design guidelines, and post-processing of manufactured parts.

Munguía, J., J. Ciurana, and C. Riba. 2009. "Neural-Network-Based Model for Build- Time Estimation in Selective Laser Sintering." *Proceedings of the Institution of Mechanical Engineers, Part B: Journal of Engineering Manufacture* 223 (8): 995–1003.

Cost assessment for rapid manufacturing (RM) is highly dependent on time estimation. Total build time dictates most indirect costs for a given part, such as labour, machine costs, and overheads. A number of parametric and empirical time estimators exist; however, they normally account for error rates between 20 and 35 per cent which are then translated to inaccurate final cost estimations. The estimator presented herein is based on the ability of artificial neural networks (ANNs) to learn and adapt to different cases, so that the developed model is capable of providing accurate estimates regardless of machine type or model. A simulation is performed with MATLAB to compare existing approaches for cost/time estimation for selective laser sintering (SLS). Error rates observed from the model range from 2 to 15 per cent, which shows the validity and robustness of the proposed method.

Neef, Andreas, Klaus Burmeister, Stefan Krempl. 2005. Vom Personal Computer zum Personal Fabricator (From Personal Computer to Personal Fabricator). Hamburg: Murmann Verlag.

Abstract unavailable

Paul, Ratnadeep, and Sam Anand. 2012. "Process Energy Analysis and Optimization in Selective Laser Sintering." *Journal of Manufacturing Systems* 31 (4): 429–37. doi:10.1016/j.jmsy.2012.07.004.

Additive manufacturing (AM) processes are increasingly being used to manufacture complex precision parts for the automotive, aerospace and medical industries. One of the popular AM processes is the selective laser sintering (SLS) process which manufactures parts by sintering metallic, polymeric and ceramic powder under the effect of laser power. The laser energy expenditure of SLS process and its correlation to the geometry of the manufactured part and the SLS process parameters, however, have not received much attention from AM/SLS researchers. This paper presents a mathematical analysis of the laser energy required for manufacturing simple parts using the SLS process. The total energy expended is calculated as a function of the total area of sintering (TAS) using a convex hull based approach and is correlated to the part geometry, slice thickness and the build orientation. The TAS and laser energy are calculated for three sample parts and the results are provided in the paper. Finally, an optimization model is presented which computes the minimal TAS and energy required for manufacturing a part using the SLS process.

Quick, Darren. 2009. "3D Bio-Printer to Create Arteries and Organs." *Gizmag*. http://www.gizmag.com/3d-bio-printer/13609/.

Abstract unavailable

Reeves, Philip. 2007. "Rapid manufacturing–Business Implementation & Global Economic Value." Econolyst Ltd, UK.

Much has been written about the benefits of additive layer manufacturing for the production of end use part otherwise known as Rapid Manufacturing (RM), as an alternative to moulding or machining or in the manufacture of increasing complex geometries. Other additive manufacturing benefits have also been discussed in the fields of materials science and mass personalisation. This paper looks beyond the scientific and physical benefits of additive manufacturing into the more practical implications of implementing RM into the main stream production environment.

The paper starts by discussing the current position of RM within the global manufacturing economy. The paper then discusses the development of a simple iterative stage methodology for RM, which can be implemented by businesses based on a six step approach. It is suggested that this could then accelerate companies through the technology selection, justification and implementation of RM, either through technology purchase or the establishment of dedicated RM supply chains.

The paper is the result of the author's engagement in both academic research projects as an industrial partner, and through experience implementing RM technologies into both end use companies and European regional technology centres.

Reeves Philip. (2008) "How the Socioeconomic Benefits of Rapid Manufacturing can Offset Technological Limitations." RAPID 2008 Conference and Exposition. Lake Buena Vista, FL: 1-12.

Abstract unavailable

Reeves, Philip. 2009. "Additive Manufacturing–A Supply Chain Wide Response to Economic Uncertainty and Environmental Sustainability." *Econolyst Limited, The Silversmiths, Crown Yard, Wirksworth, Derbyshire, DE4 4ET, UK.*

In this paper the author will review some of the current commercial applications of Additive Layer Manufacturing (ALM) and the business benefits associated with technology adoption. The paper will review applications such as Rapid Tooling, where ALM processes are being used to make fully dense tool cavity inserts with highly efficient heating and cooling channels. This approach has been proven to have clear down-stream economic benefits within the supply chain, resulting in reduced cycle times, improved moulding quality and a lower carbon footprint.

The paper will also address how ALM is being used as a sustainable alternative to subtractive machining in the production of high buy-to-fly ratio parts, and how different Design-For- Manufacturing (DFM) rules associated with ALM, are being exploited to manufacture lighter weight, energy efficient products with less raw material. The paper concludes with a look into the future, possibly into a 'tool-less' society, where consumer products are printed to order, using the consumers own design data as-and-when they are needed, using either a globally distributed just-in-time supply chain or inversely manufacture within the consumers own home.

Rickenbacher, L., A. Spierings, and K. Wegener. 2013. "An Integrated Cost-Model for Selective Laser Melting (SLM)." *Rapid Prototyping Journal* 19 (3): 208–14.

Purpose – The integration of additive manufacturing (AM) processes into a production environment requires a cost-model that allows the precise estimation of the total cost per part, although the part might be produced in the same build job together with other parts of different sizes, complexities and quantities. Several cost-models have been proposed in the past, but most of them are not able to calculate the costs for each single part in a mixed build job or are not suitable for Selective Laser Melting (SLM). The purpose of this paper is to develop a cost model, including all pre- and post- processing steps linked to SLM.

Design/methodology/approach – Based on collected data and the generic cost model of Alexander et al., an adapted model was developed for the SLM process including all required pre- and post- processes. Each process was analysed and modelled in detail, allowing an evaluation of the influences of the different geometries on the cost of each part.

Findings – By simultaneously building up multiple parts, the manufacturing as well as the set-up time and therefore the total cost per part can be significantly reduced. In the presented case study a cost reduction of 41 per cent can be achieved in average.

Originality/value – Using different cost allocation algorithms, the developed cost model enables a precise determination of total cost per part avoiding that any geometry is preferred in simultaneous manufacture. This helps to optimize build jobs and to manufacture SLM parts more economically by pooling parts from different projects, whereas the cost per part can still be precisely determined.

Ruffo, M., and R. Hague. 2007. "Cost Estimation for Rapid Manufacturing-Simultaneous Production of Mixed Components Using Laser Sintering." *Proceedings of the Institution of Mechanical Engineers, Part B: Journal of Engineering Manufacture* 221 (11): 1585–91.

Rapid manufacturing (RM) is a production method able to build components by adding material layer by layer, and it thus allows the elimination of tooling from the production chain. For this reason, RM enables a cost-efficient production of low-volume components favouring the customization strategy. Previous work has been developed on costing methodologies applicable to RM, but it was limited to the scenario of the production of copies of the same part. In reality, RM enables the production of different components simultaneously, and thus a smart mix of components in the same machine can achieve an enhanced cost reduction. This paper details this concept by proposing mathematical models for the assignment of the full production cost into each single product and by validating through a case study. This paper extends previous work on RM costing by adding the scenario of simultaneous production of different parts.

Ruffo, M., C. Tuck, and R. Hague. 2006a. "Cost Estimation for Rapid Manufacturing- Laser Sintering Production for Low to Medium Volumes." *Proceedings of the Institution of Mechanical Engineers, Part B: Journal of Engineering Manufacture* 220 (9): 1417–27.

Rapid manufacturing (RM) is a modern production method based on layer by layer manufacturing directly from a three-dimensional computer-

aided design model. The lack of tooling makes RM economically suitable for low and medium production volumes. A comparison with traditional manufacturing processes is important; in particular, cost comparison. Cost is usually the key point for decision making, with break-even points for different manufacturing technologies being the dominant information for decision makers. Cost models used for traditional production methodologies focus on material and labour costs, while modern automated manufacturing processes need cost models that are able to consider the high impact of investments and overheads. Previous work on laser sintering costing was developed in 2003. This current work presents advances and discussions on the limits of the previous work through direct comparison. A new cost model for laser sintering is then proposed. The model leads to graph profiles that are typical for layer-manufacturing processes. The evolution of cost models and the indirect cost significance in modern costing representation is shown finally.

Ruffo, M., C. Tuck, and R. Hague. 2006b. "Empirical Laser Sintering Time Estimator for Duraform PA." *International Journal of Production Research* 44 (23): 5131–46.

The paper presents work on the development of a build-time estimator for rapid manufacturing. A time estimator is required to develop a comprehensive costing tool for rapid manufacturing. An empirical method was used to estimate build times using both simulated and actual builds for a laser sintering machine. The estimator presented herein is based upon object geometry and, therefore, the fundamental data driving the model are obtainable from current three-dimensional computer-aided design models. The aim is to define a model describing the build times for a laser sintering machine either for single or multiple objects.

Ruffo, M., C. Tuck, and R. Hague. 2007. "Make or Buy Analysis for Rapid Manufacturing." *Rapid Prototyping Journal* 13 (1): 23–29.

Purpose – The purpose of this paper is to outline how rapid manufacturing (RM) could influence the decision-making process for managers involved in make or buy decisions.

Design/methodology/approach – A literature review on make or buy issues has been carried out and the results of which have been distilled into a number of qualitative considerations. These considerations have been formed into three possible make or buy scenarios: the firm has no experience of rapid prototyping (RP) or RM; the firm already has an RP department; and the firm already has an RM function. In order to analyse the decision further a quantitative approach has been taken, mainly adapted to the last scenario but applicable also to the second scenario. Here, manufacturing cost data has been directly compared with price information from two current RP bureaus. The differences between RM cost and RP price have been studied.

Findings – Strategically, the points analysed were in favour of the make option. Economically, the lack of dedicated RM bureaus and the consequent use of RP costing has further pushed the make or buy decision in favour of make.

Originality/value – There is a lack of work on the implementation of RM as a mainstream manufacturing process. Existing knowledge has begun to look at the use and costs of RM, however, this paper highlights the lack of dedicated RM providers.

Senyana, Lionel Nduwayezu. 2011. "Environmental Impact Comparison of Distributed and Centralized Manufacturing Scenarios". Rochester Institute of Technology.

Centralized manufacturing and distributed manufacturing are two fundamentally different methods for producing components. This work describes a centralized manufacturing scenario in which parts are produced via forging and finish machining at one central location and are then shipped to the end user. The distributed manufacturing model involves a scenario in which an additive manufacturing process (Electron Beam Melting) is used to produce parts to near net shape with minimal finish machining. Because the process doesn't require molds or dies, production can take place in small production quantities "on demand" at job shops located close to the end user with little transportation. In other words, parts are not produced until they are needed. This is in stark contrast to the centralized model where large quantities of parts are produced and then distributed at a later date when needed from warehouses. The aim of this thesis is to compare the environmental impact of these two different production approaches under a variety of conditions. The SimaPro software package has been used to model both approaches with input from the user involving part size, amount of finish machining, transportation distances, mode of transportation, production quantities, etc. Results from simulation models indicate that at small production quantities, the environmental impact of forging die production dominates the centralized manufacturing model. As production quantity increases, finish machining begins to dominate the environmental impact. Despite the large transportation distances involved, the transportation distance and mode of transportation actually have relatively little impact on overall environmental impact compared with other factors. Regardless of the production scenario being evaluated, the distributed manufacturing approach had less environmental impact. The production of titanium powder as the raw material contributed the majority of environmental impact for this approach. Although this work examines environmental impact, it does not consider the cost of producing a part. It should be pointed out, however, that the distributed manufacturing approach could someday have a profound effect on supply chain management for replacement parts by reducing or eliminating the need for warehouses along with associated inventory carrying costs, product obsolescence costs, heating and cooling energy, etc.

Sreenivasan, R., and D.L. Bourell. 2009. "Sustainability Study in Selective Laser Sintering – An Energy Perspective." In 20th Annual International Solid Freeform Fabrication Symposium–An Additive Manufacturing Conference, Austin/TX/USA, 3rd–5th August. Austin, TX.

This paper presents a sustainability analysis of Selective Laser Sintering (SLS) from an energy standpoint. Data of electrical power consumed by the system over an entire build were acquired using a LabVIEW 8.6 circuit. The power drawn by individual subsystems were also measured, and an energy balance was performed. These data were then used to arrive at a Total Energy Indicator of the process with the help of a specific type of Environmental and Resource Management Data (ERMD) known as Eco-Indicators, which indicates the level of sustainability of the process.

Sreenivasan, R., A. Goel, and D.L. Bourell. 2010. "Sustainability Issues in Laser-Based Additive Manufacturing." *Physics Procedia* 5 (January): 81–90. doi:10.1016/j.phpro.2010.08.124.

Sustainability is a consideration of resource utilization without depletion or adverse environmental impact. In manufacturing, important sustainability issues include energy consumption, waste generation, water usage and the environmental impact of the manufactured part in service. This paper deals with three aspects of sustainability as it applies to additive manufacturing. First is a review of the research needs for energy and sustainability as applied to additive manufacturing based on the 2009 Roadmap for Additive Manufacturing Workshop. The second part is an energy assessment for selective laser sintering (SLS) of polymers. Using polyamide powder in a 3D Systems Vanguard HiQ Sinterstation, energy loss during a build was measured due to the chamber heaters, the roller mechanism, the piston elevators and the laser. This accounted for 95% of the total energy consumption. An overall energy assessment was accomplished using eco-indicators. The last topic is electrochemical deposition of porous SLS non-polymeric preforms. The goal is to reduce energy consumption in SLS of non-polymeric materials. The approach was to mix a transient binder with the material, to create an SLS green part, to convert the binder, and then to remove the open, connected porosity and to densify the part by chemical deposition at room temperature within the pore network. The model system was silicon carbide powder mixed with a phenolic transient binder coupled with electrolytic deposition of nickel. Deposition was facilitated by inserting a conductive graphite cathode in the part center to draw the positive nickel ions through the interconnected porous network and to deposit them on the pore walls.

Stoneman, Paul. The Economics of Technological Diffusion. 2002. Oxford: Blackwell.

This book presents a detailed overview of the economics of technological diffusion in all its various dimensions. Topics covered include:

- *Game-theoretic approaches to the modelling of technological change*
- *Finance and technological change*
- *Technological change in international trade.*

Telenko, Cassandra, and Carolyn Conner Seepersad. 2010. "Assessing Energy Requirements and Material Flows of Selective Laser Sintering of Nylon Parts." In *21st Annual International Solid Freeform Fabrication Symposium–An Additive Manufacturing Conference, Austin/TX/USA, 6th–8th August*, 8–10. Austin, TX.

Selective laser sintering (SLS) is a prominent technology for rapid manufacturing (RM) of functional parts. SLS and competitive RM technologies are generally assumed to be more environmentally sustainable than conventional manufacturing methods because the additive process minimizes tooling, material waste, and chemical fluids. A thorough life cycle analysis (LCA) of the environmental impacts of SLS has yet to be published. This study focuses on a section of the SLS part life-cycle. It tracks the nylon powder material flows from the extraction and synthesis of the material to SLS part production. Basic material properties and environmental effects are reported. Estimates of material waste and energy use are also reported and compared with those of injection molding.

Telenko, Cassandra, and Carolyn Conner Seepersad. 2011. "A Comparative Evaluation of Energy Consumption of Selective Laser Sintering and Injection Molding of Nylon Parts." *Rapid Prototyping J* 18: 472–81.

Additive manufacturing is often advocated as a sustainable alternative to competing manufacturing technologies. This research study focuses on estimating and comparing the energy consumption required for different production volumes of nylon parts using either selective laser sintering (SLS) or injection molding (IM). For IM & SLS, energy consumption is estimated for nylon material refinement and part fabrication. For IM, energy consumption is also estimated for manufacturing the injection molds and refining their metal feedstock. A paintball gun handle serves as a representative part for calculating and normalizing material flows and processing times. For different sets of assumptions, cross-over production volumes are calculated, at which the per-part energy consumption of the two processes is equivalent. These energy-based cross-over production volumes are compared to similar economic cross-over production volumes available in the literature.

Telenko, Cassandra, and Carolyn Conner Seepersad. 2012. "A Comparison of the Energy Efficiency of Selective Laser Sintering and Injection Molding of Nylon Parts." *Rapid Prototyping Journal* 18 (6): 472–81.

Purpose – The purpose of this paper is to evaluate the energy consumed to fabricate nylon parts using selective laser sintering (SLS) and to compare it with the energy consumed for injection molding (IM) the same parts.

Design/methodology/approach – Estimates of energy consumption include the energy consumed for nylon material refinement, adjusted for SLS and IM process yields. Estimates also include the energy consumed by the SLS and IM equipment for part fabrication and the energy consumed to machine the injection mold and refine the metal feedstock required to fabricate it. A representative part is used to size the injection mold and to quantify throughput for the SLS machine per build.

Findings – Although SLS uses significantly more energy than IM during part fabrication, this energy consumption is partially offset by the energy consumption associated with production of the injection mold. As a result, the energy consumed per part for IM decreases with the number of parts fabricated while the energy consumed per part for SLS remains relatively constant as long as builds are packed efficiently. The crossover production volume, at which IM and SLS consume equivalent amounts of energy per part, ranges from 50 to 300 representative parts, depending on the choice of mold plate material.

Research limitations/implications – The research is limited to material refinement and part fabrication and does not consider other aspects of the life cycle, such as waste disposal, distributed 2 manufacturing, transportation, recycling or use. Also, the crossover volumes are specific to the representative part and are expected to vary with part geometry.

Originality/value – The results of this comparative study of SLS and IM energy consumption indicate that manufacturers can save energy using SLS for parts with small production volumes. The comparatively large amounts of nylon material waste and energy consumption during fabrication make it inefficient, from an energy perspective, to use SLS for higher production volumes. The crossover production volume depends on the geometry of the part and the choice of material for the mold.

Thomas, Douglas. 2013. *Economics of the U.S. Additive Manufacturing Industry*. NIST Special Publication 1163. Gaithersburg, MD: U.S. Dept. of Commerce, National Institute of Standards and Technology.

There is a general concern that the U.S. manufacturing industry has lost competitiveness with other nations. Additive manufacturing may provide an important opportunity for advancing U.S. manufacturing while maintaining and advancing U.S. innovation. Additive manufacturing is a relatively new process where material is joined together layer by layer to make objects from three- dimensional models as opposed to conventional methods where material is removed. The U.S. is currently a major user of additive manufacturing technology and the primary producer of additive manufacturing systems. Globally, an estimated $642.6 million in revenue was collected for additive manufactured goods, with the U.S. accounting for an estimated $246.1 million or 38.3% of global production in 2011. Change agents for the additive manufacturing industry can focus their efforts on three primary areas to advance this technology: cost reduction, accelerating the realization of benefits, and increasing the benefits of additive manufacturing. Significant impact on these areas may be achieved through reduction in the cost of additive manufacturing system utilization, material costs, and facilitating the production of large products. There is also a need for a standardized model for cost categorization and product quality and reliability testing.

Tuck, Christopher, Richard Hague, and Neil Burns. 2007. "Rapid Manufacturing: Impact on Supply Chain Methodologies and Practice." *International Journal of Services and Operations Management* 3 (1): 1–22.

This paper demonstrates the use of Rapid Manufacturing (RM) as the enabling technology for flexible manufacturing in a number of industrial sectors. This paper discusses the evolution of Rapid Prototyping (RP) to

RM and the current issues that require further research for the successful integration of this technology within manufacturing companies. The use of RM will have particular impact on supply chain management paradigms such as lean and agile and has particular strategic fit with mass customisation. The effect of RM will have on these paradigms is discussed and confirmed with example cases from automotive production, motor sport and medical devices industries. In conclusion, RM has already been shown in the three cases to offer benefits, particularly where fast reconfiguration of the manufacturing process is required and with the production of customised components.

University of San Francisco. Walmart: Keys to Successful Supply Chain Management.<http://www.usanfranonline.com/resourcessuccessful-supply-chain-management/#.U5IDQfldXzg>

Abstract unavailable

Vasquez, Mike. 2009. "Economic and Technological Advantages of Using High Speed Sintering as a Rapid Manufacturing Alternative in Footwear Applications." Massachusetts Institute of Technology.

Rapid manufacturing is a family of technologies that employ additive layer deposition techniques to construct parts from computer based design models.[2] These parts can then be used as prototypes or finished goods. One type of rapid manufacturing technology, Selective Laser Sintering, only allows for a point-by-point sintering process to construct the 3D representations of CAD models. This makes for long processing periods and is ineffective for high volume manufacturing. However, a new process called high-speed sintering uses infrared energy to 'flash' the polymer powder at multiple points making the layer deposition process much more time efficient. In effect each infusion of energy results in an entire layer being constructed rather than a single point. One of the first industrial applications for this technique is in performance footwear manufacturing. New Balance, a Boston based shoe and apparel company, in collaboration with Loughborough University has an interest in exploring the technology for low volume parts manufacturing as well as personalized footwear. High speed sintering has the potential to replace injection molding for specific footwear and non-footwear applications.

This technology has several key advantages over injection molding including the ability to build complex geometries that would be impossible with injection molding. Also as the technology continues to evolve new materials could improve the mechanical performance of finished parts. Nevertheless, as with commercializing any new technology identifying a cost effective implementation route is a pivotal step. (cont.) This project addressed this concern by thoroughly investigating the current and potential state of high speed sintering. The manufacture of a New Balance shoe part using both high speed sintering and injection molding was directly compared. Several factors including time to manufacture and cost were investigated.

Verma, Anoop, and Rahul Rai. 2013. "Energy Efficient Modeling and Optimization of Additive Manufacturing Processes." In *24th Annual International Solid Freeform Fabrication Symposium–An Additive Manufacturing Conference, Austin/TX/USA*. Austin, TX.

Additive manufacturing (AM) is a leading technology in various industries including medical and aerospace for prototype and functional part fabrication. Despite being environmentally conscious, avenues pertaining to further reducing the impact of AM on the environment exist. Material wastage and energy consumption are two major concerns of the process that requires immediate attention. In this research, a multi-step optimization enabling additive manufacturing process towards energy efficiency is developed. Process objectives such as material waste and energy consumption are minimized both in part and layer domain. Numerous examples are presented to demonstrate the applicability of the developed approach. The models formulated here for selective laser sintering (SLS) process can be easily extended to other additive manufacturing technologies.

Walter, Manfred, Jan Holmström, H. Tuomi, and H. Yrjölä. 2004. "Rapid Manufacturing and Its Impact on Supply Chain Management." In *Proceedings of the Logistics Research Network Annual Conference*, 9–10.

Suppliers of spare parts suffer from high inventory and distribution costs in many industries. Original Equipment Manufacturers (OEMs) have attempted to reduce these supply chain costs by cutting production lead-times, batch constraints and delivery lead-times. The emphasis in supply chain management has been towards increased inventory turnover.

Today, rapid manufacturing technologies – the ability to produce parts on demand without the need for tooling and setup – has the potential to become the basis for new solutions in supply chain management. This paper presents new supply chain solutions made possible by both the centralised and decentralised applications of rapid manufacturing. A decision-support model is outlined to help supply chain managers better capture emergent business opportunities arising from rapid manufacturing technology.

The logistical problems of the spare parts business in the aircraft industry are used as an example due to the high technical and logistical requirements involved. The applications and benefits of rapid manufacturing technologies in the supply chain for aircraft spare parts are presented.

Wohlers, Terry. "Wohlers Report 2012: Additive Manufacturing and 3D Printing State of the Industry." Wohlers Associates, Inc. 2012.

Abstract unavailable.

Wohlers, Terry. "Wohlers Report 2014: Additive Manufacturing and 3D Printing State of the Industry." Wohlers Associates, Inc. 2014.

Abstract unavailable.

Young, Son K. "A Cost Estimation Model for Advanced Manufacturing Systems." *International Journal of Production Research*. 1991. 29(3): 441-452.

As manufacturers continue to automate their factories, they discover that existing cost measures should be updated. Much of the existing literature has discussed the 'why's' but there is little about the 'how's.' This paper expands the cost concept to include quality and flexibility because they are critical factors for performance evaluation and project justification of advanced manufacturing systems. Then, a quantitative method of estimating the cost elements is illustrated. Finally, various approaches to collecting parametric values of the cost model and applications of the cost model are presented.

Zhai, Yun. 2012. "Early Cost Estimation for Additive Manufacture". Cranfield University.

Additive Manufacture (AM) is a novel manufacturing method; it is a process of forming components by adding materials. Owing to material saving and manufacturing cost saving, more and more research has been focused on metal AM technologies. WAAM is one AM technology, using arc as the heat sources and wire as the material to create parts with weld beads on a layer-by-layer basis. The process can produce components in a wide range of materials, including aluminum, titanium and steel. High deposition rate, material saving and elimination of tooling cost are critical characteristics of the process. Cost estimation is important for all companies. The estimated results can be used as a datum to create a quote for customers or evaluate a quote from suppliers, an important consideration for the application of WAAM is its cost effectiveness compared with traditional manufacture methods. The aim of this research is to find a way to develop a cost estimating method capable of providing manufacturing cost comparison of WAAM with CNC. A cost estimation model for CNC machining has been developed. A process planning approach for WAAM was also defined as part of this research. An Excel calculation spreadsheet was also built and it can be easily used to estimate and compare manufacture cost of WAAM with CNC. Using the method developed in this research, the cost driver analysis of WAAM has been made. The result shows that reduced material cost is the biggest cost driver in WAAM. The cost comparison of WAAM and CNC also has been made and the results show that with the increase of buy-to-fly ratio WAAM is more economical than CNC machining.

Zhang, Y, and A Bernard. 2014. "Generic Build Time Estimation Model for Parts Produced by SLS." In High Value Manufacturing: Advanced Research in Virtual and Rapid Prototyping: Proceedings of the 6th International Conference on Advanced Research in Virtual and Rapid Prototyping, Leiria, Portugal, 1-5 October, 2013.

Rapid Prototyping (RP) has evolved into Additive Manufacturing (AM) and plays an important role in numerous application domains. Cost and lead time of AM become significant factors affecting the comparison between AM and other traditional processes. The accuracy of build time estimation directly affects the cost estimation for AM production. This paper introduces an analytical method to build time estimation for parts, which takes real AM production context that was usually neglected by former models into consideration. To illustrate the proposed method, an analytical generic build time estimation model is constructed for SLS process with a simple calculation example. The results reflect the importance of production context for the build time estimation.

End Notes

[1] Economist. "Printing Body Parts: Making a Bit of Me." <http: //www.economist.com /node/15543683>

[2] Quick 2009. "3D Bio-printer to Create Arteries and Organs." < http:// www.gizmag.com/3d-bio- printer/13609/>

[3] Wohlers, Terry. "Wohlers Report 2012: Additive Manufacturing and 3D Printing State of the Industry." Wohlers Associates, Inc. 2012.

[4] Wohlers, Terry. "Wohlers Report 2014: Additive Manufacturing and 3D Printing State of the Industry."Wohlers Associates, Inc. 2014: 129.

[5] This value is calculated with the assumption that the U.S. share of additive manufacturing systems sold equates to the share of products produced using additive manufacturing systems. The share of additive manufacturing systems is available in Wohlers, Terry. "Wohlers Report 2012: Additive Manufacturing and 3D Printing State of the Industry." Wohlers Associates, Inc. 2012: 134.

[6] Wohlers, Terry. "Wohlers Report 2012: Additive Manufacturing and 3D Printing State of the Industry."Wohlers Associates, Inc. 2012: 130.

[7] Neef, Andreas, Klaus Burmeister, Stefan Krempl. 2005. Vom Personal Computer zum Personal Fabricator (From Personal Computer to Personal Fabricator). Hamburg: Murmann Verlag.

[8] Neef, Andreas, Klaus Burmeister, Stefan Krempl. 2005. Vom Personal Computer zum Personal Fabricator (From Personal Computer to Personal Fabricator). Hamburg: Murmann Verlag.

[9] Baumers, Martin. "Economic Aspects of Additive Manufacturing: Benefits, Costs, and Energy Consumption." 2012. Doctoral Thesis. Loughborough University.

[10] 3D Systems purchased Z Corporation in 2012. Stratasys merged with Objet in 2012 and is now incorporated in Israel.

[11] Thomas, Douglas S. Economics of the U.S. Additive Manufacturing Industry. NIST Special Publication 1163. 2013. <http://www.nist.gov/manuscript-publication-search.cfm?pub_id= 913515>

[12] Wohlers, Terry. "Wohlers Report 2012: Additive Manufacturing and 3D Printing State of the Industry." Wohlers Associates, Inc. 2012.

[92] Wohlers, Terry. "Wohlers Report 2012: Additive Manufacturing and 3D Printing State of the Industry." Wohlers Associates, Inc. 2012.

[14] Young, Son K. "A Cost Estimation Model for Advanced Manufacturing Systems." *International Journal of Production Research*. 1991. 29(3): 441-452.

[15] For this report, medium- and high-tech manufacturing includes NAICS 333 through 336, which includes machinery, computer, electronic product, electrical equipment, and transportation equipment manufacturing.

[16] It is assumed that the cost of holding inventory is 25% of the value of the inventory.

[17] Reeves P. (2008) "How the Socioeconomic Benefits of Rapid Manufacturing can Offset Technological Limitations." RAPID 2008 Conference and Exposition. Lake Buena Vista, FL: 1-12.

[18] Walter, Manfred, Jan Holmstrom and Hannu Yrjola. “Rapid Manufacturing and its Impact on Supply Chain Management.” Logistics Research Network Annual Conference. September 9-10, 2004. Dublin, Ireland.

[19] Neef, Andreas, Klaus Burmeister, Stefan Krempl. 2005. Vom Personal Computer zum Personal Fabricator (From Personal Computer to Personal Fabricator). Hamburg: Murmann Verlag.

[20] Neef, Andreas, Klaus Burmeister, Stefan Krempl. 2005. Vom Personal Computer zum Personal Fabricator (From Personal Computer to Personal Fabricator). Hamburg: Murmann Verlag.

[21] Huang, Samuel H., Peng Liu, Abhiram Mokasdar. 2013 “Additive Manufacturing and Its Societal Impact: A Literature Review.” International Journal of Advanced Manufacturing Technology. 67: 1191-1203.

[22] Holmstrom, Jan, Jouni Partanen, Jukka Tuomi, and Manfred Walter. “Rapid Manufacturing in the Spare Parts Supply Chain: Alternative Approaches to Capacity Deployment.” Journal of Manufacturing Technology Management. 2010. 21(6) 687-697.

[23] Khajavi, Siavash H., Jouni Partanen, Jan Holmstrom. 2014 “Additive Manufacturing in the Spare Parts Supply Chain.” Computers in Industry. 65: 50-63.

[24] Holmström, Jan, Jouni Partanen, Jukka Tuomi, and Manfred Walter. 2010. “Rapid Manufacturing in the Spare Parts Supply Chain: Alternative Approaches to Capacity Deployment.” Journal of Manufacturing Technology. 21(6): 687-697.

[25] University of San Francisco. Walmart: Keys to Successful Supply Chain Management. <http://www.usanfranonline.com/resources chain-management/#.U5IDQfldXzg>

[26] Atzeni, Eleonora and Alessandro Salmi. (2012) “Economics of Additive Manufacturing for End-Usable Metal Parts.” International Journal of Advanced manufacturing Technology. 62: 1147-1155.

[27] Lindemann C., U. Jahnke, M. Moi, and R. Koch. “Analyzing Product Lifecycle Costs for a Better Understanding of Cost Drivers in Additive Manufacturing.” Proceedings of the 2012 Solid Freeform Fabrication Symposium. <http://utwired.engr.utexas.edu/lff/symposium /proceedings Archive/pubs/Manuscripts/2012/2012-12- Lindemann.pdf>

[28] Baumers, Martin. “Economic Aspects of Additive Manufacturing: Benefits, Costs, and Energy Consumption.” 2012. Doctoral Thesis. Loughborough University.

[29] Stoneman, Paul. The Economics of Technological Diffusion. 2002. Oxford: Blackwell.

[30] Atzeni, Eleonora, Luca Iuliano, Paolo Minetola, and Alessandro Salmi. 2010. "Redesign and Cost Estimation of Rapid Manufactured Plastic Parts." *Rapid Prototyping Journal* 16 (5): 308–17.

[31] Hopkinson, Neil, and Phill M. Dickens. "Analysis of Rapid Manufacturing – Using Layer Manufacturing Processes for Production." Proceedings of the Institution of Mechanical Engineers, Part C : Journal of Mechanical Engineering Science. 2003. 217(C1): 31-39. <https://dspace.lboro.ac.uk/dspace- jspui/handle/2134/3561>

[32] Ruffo, M, Christopher Tuck, Richard J.M. Hague. "Cost Estimation for Rapid Manufacturing – Laser Sintering Production for Low to Medium Volumes." Proceedings of the Institution of Mechanical Engineers, Part B: Journal of Engineering Manufacture. 2006. 1417-1427. <https:// dspace.lboro.ac.uk/dspace-jspui/handle/2134/4680>

[33] Wohlers, Terry. "Wohlers Report 2012: Additive Manufacturing and 3D Printing State of the Industry." Wohlers Associates, Inc. 2012.

[34] Ibid

[35] The build envelope is the maximum area for part production in an additive manufacturing system.

[36] Ruffo, Massimiliano, Christopher Tuck, and Richard Hague. 2006. "Empirical Laser Sintering Time Estimator for Duraform PA." *International Journal of Production Research* 44 (23): 5131–46.

[37] Campbell, I., J. Combrinck, D. De Beer, and L. Barnard. 2008. "Stereolithography Build Time Estimation Based on Volumetric Calculations. Rapid Prototyping Journal. 14(5): 271-279.

[38] Di Angelo, Luca, and Paolo Di Stefano. 2011. "A Neural Network-Based Build Time Estimator for Layer Manufactured Objects." *International Journal of Advanced Manufacturing Technology* 57 (1-4): 215–24. doi:10.1007/s00170-011-3284-8.

[39] Hopkinson, Neil, and Phill M. Dickens. "Analysis of Rapid Manufacturing – Using Layer Manufacturing Processes for Production." Proceedings of the Institution of Mechanical Engineers, Part C : Journal of Mechanical Engineering Science. 2003. 217(C1): 31-39. <https://dspace.lboro.ac.uk/dspace- jspui/handle/2134/3561>

[40] Baumers, Martin. "Economic Aspects of Additive Manufacturing: Benefits, Costs, and Energy Consumption." 2012. Doctoral Thesis. Loughborough University.

[41] Morrow, W.R., H. Qi, I. Kim, J. Mazumder, and S.J. Skerlos. 2007. "Environmental Aspects of Laser - Based and Conventional Tool and Die Manufacturing." *Journal of Cleaner Production* 15 (10): 932–43. doi:10.1016/j.jclepro.2005.11.030.

[42] Mognol, Pascal, Denis Lepicart, and Nicolas Perry. 2006. "Rapid Prototyping: Energy and Environment in the Spotlight." *Rapid Prototyping Journal* 12 (1): 26–34. doi:10.1108/13552540610637246.

[43] Telenko, Cassandra, and Carolyn Conner Seepersad. 2012. "A Comparison of the Energy Efficiency of Selective Laser Sintering and Injection Molding of Nylon Parts." *Rapid Prototyping Journal* 18 (6): 472–81.

[44] Sreenivasan, R., and D.L. Bourell. 2009. "Sustainability Study in Selective Laser Sintering – An Energy Perspective." In *20th Annual International Solid Freeform Fabrication Symposium–An Additive Manufacturing Conference, Austin/TX/USA, 3rd–5th August*. Austin, TX.

[45] Baumers, Martin. "Economic Aspects of Additive Manufacturing: Benefits, Costs, and Energy Consumption." 2012. Doctoral Thesis. Loughborough University.

[46] Doubrovski, Zjenja, Jouke C. Verlinden, and Jo M.P. Geraedts. "Optimal Design for Additive Manufacturing: Opportunities and Challenges." Proceedings of the ASME 2011 International Design Engineering Technical Conferences and Computers and Information in Engineering Conference. August 29-31, 2011. Washington DC.

[47] Ruffo, M, Christopher Tuck, Richard J.M. Hague. "Cost Estimation for Rapid Manufacturing – Laser Sintering Production for Low to Medium Volumes." Proceedings of the Institution of Mechanical Engineers, Part B: Journal of Engineering Manufacture. 2006. 1417-1427.
<https://dspace.lboro.ac.uk/dspace-jspui/handle/2134/4680>

[48] Hopkinson, Neil, and Phill M. Dickens. "Analysis of Rapid Manufacturing – Using Layer Manufacturing Processes for Production." Proceedings of the Institution of Mechanical Engineers, Part C : Journal of Mechanical Engineering Science. 2003. 217(C1): 31-39. <https://dspace.lboro.ac.uk/dspace- jspui/handle/2134/3561>

[49] Baumers, Martin. "Economic Aspects of Additive Manufacturing: Benefits, Costs, and Energy Consumption." 2012. Doctoral Thesis. Loughborough University.

[50] Allen, Jeff. 2006. "An Investigation into the Comparative Costs of Additive Manufacture vs. Machine from Solid for Aero Engine Parts." In Cost Effective Manufacture via Net-Shape Processing, 17-1 – 17-10. Meeting

Proceedings RTO-MP-AVT-139. Paper 17. DTIC Document. <http://www.rto.nato.int/abstracts.asp>

[51] Calculated from data in the Annual Survey of Manufactures and the Quarterly survey of plant capacity utilization.

[52] Kim, Bowon. "Supply Chain Management: A Learning Perspective." Korea Advanced Institute of Science and Technology. Coursera Lecture 1-2.

[53] Kim, Bowon and Chulsoon Park. (2013). "Firms' Integrating Efforts to Mitigate the Tradeoff Between Controllability and Flexibility." International Journal of Production Research. 51(4): 1258-1278.

[54] Wohlers, Terry. "Wohlers Report 2012: Additive Manufacturing and 3D Printing State of the Industry." Wohlers Associates, Inc. 2012.

[55] Ibid

[56] Ibid.

[57] Thomas, Douglas. 2013. *Economics of the U.S. Additive Manufacturing Industry*. NIST Special Publication 1163. Gaithersburg, MD: U.S. Dept. of Commerce, National Institute of Standards and Technology.

[58] Wohlers, Terry. "Wohlers Report 2012: Additive Manufacturing and 3D Printing State of the Industry." Wohlers Associates, Inc. 2012.

[59] Mansfield, Edwin. *Innovation, Technology and the Economy: Selected Essays of Edwin Mansfield.* Economists of the Twentieth Century Series (Brookfield, VT: 1995, E. Elgar).

[60] Chapman, Robert. "Benefits and Costs of Research: A Case Study of Construction Systems Integration and Automation Technologies in Commercial Buildings." NISTIR 6763. December 2001. National Institute of Standards and Technology.

[61] The price was adjusted using the Consumer Price Index for all consumers for all areas from the Bureau of Labor Statistics. This adjustment, likely, underestimates the degree of price deflation, as it does not account for quality and productivity improvements specific to these systems. Unfortunately, there is not a price index that accounts for these issues.

In: Additive Manufacturing
Editor: Felipe Brewer
ISBN: 978-1-63483-364-6

Chapter 2

ECONOMICS OF THE U.S. ADDITIVE MANUFACTURING INDUSTRY*

Douglas S. Thomas

ABSTRACT

There is a general concern that the U.S. manufacturing industry has lost competitiveness with other nations. Additive manufacturing may provide an important opportunity for advancing U.S. manufacturing while maintaining and advancing U.S. innovation. Additive manufacturing is a relatively new process where material is joined together layer by layer to make objects from three-dimensional models as opposed to conventional methods where material is removed. The U.S. is currently a major user of additive manufacturing technology and the primary producer of additive manufacturing systems. Globally, an estimated $642.6 million in revenue was collected for additive manufactured goods, with the U.S. accounting for an estimated $246.1 million or 38.3% of global production in 2011. Change agents for the additive manufacturing industry can focus their efforts on three primary areas to advance this technology: cost reduction, accelerating the realization of benefits, and increasing the benefits of additive manufacturing. Significant impact on these areas may be achieved through reduction in the cost of additive manufacturing system utilization, material costs, and facilitating the production of large

* This is an edited, reformatted and augmented version of NIST Special Publication 1163, issued by the National Institute of Standards and Technology, August 2013.

products. There is also a need for a standardized model for cost categorization and product quality and reliability testing.

PREFACE

This study was conducted by the Applied Economics Office in the Engineering Laboratory at the National Institute of Standards and Technology. The study provides aggregate manufacturing industry data and industry subsector data to develop a quantitative depiction of the U.S. additive manufacturing industry.

1. INTRODUCTION

1.1. Background

In 2010, the world produced approximately $10.2 trillion in manufacturing value added, according to United Nations Statistics Division (UNSD) data. The U.S. produced approximately 18% of these goods, making it the second largest manufacturing nation in the world, down from being the largest in 2009. Many products and parts made by the industry are produced by taking pieces of raw material and cutting away sections to create the desired part; however, a relatively new process called additive manufacturing is beginning to take hold where material is aggregated together rather than cut away. Additive manufacturing is the process of joining materials to make objects from three-dimensional (3D) models layer by layer as opposed to subtractive methods that remove material. The terms additive manufacturing and 3D printing tend to be used interchangeably to describe the same approach to fabricating parts. This technology is used to produce models, prototypes, patterns, components, and parts using a variety of materials including plastic, metal, ceramics, glass, and composites. Products with moving parts can be printed such that the pieces are already assembled. Technological advances have even resulted in a 3D-Bio-printer that one day might create body parts on demand.[1,2]

Additive manufacturing is used by multiple industry subsectors, including motor vehicles, aerospace, machinery, electronics, and medical products.[3] This technology dates back to the 1980's with the development of stereolithography, which is a process that solidifies layers of liquid polymer using a laser. The first additive manufacturing system available was the SLA-1 by 3D Systems. Technologies that enabled the advancement of additive manufacturing were the desktop computer and the availability of industrial lasers.

Although additive manufacturing allows the manufacture of increasingly complex parts, the slow print speed of additive manufacturing systems limits their use for mass production. 3D scanning technologies have enabled the replication of real objects without using molds, which can be difficult and expensive. As the costs of additive manufacturing systems decrease, this technology may change the way that consumers interact with producers. The customization of products will require increased data collection from the end user. Additionally, an inexpensive 3D printer allows the end user to produce polymer-based products in their own home or office. Currently, there are a number of systems that are within the budget of the average consumer.

1.2. Purpose

Additive manufacturing technology opens up new opportunities for the economy and society. It can facilitate the production of strong light-weight products for the aerospace industry and it allows designs that were not possible with previous manufacturing techniques. It may revolutionize medicine with biomanufacturing. This technology has the potential to increase the well-being of U.S. citizens and improve energy efficiency in ground and air transportation. However, the adoption and diffusion of this new technology is not instantaneous. With any new technology, new standards, knowledge, and infrastructure are required to facilitate its use. Organizations such as the National Institute of Standards and Technology can enable the development of these items; thus, it is important to understand the size and extent of the additive manufacturing industry. Although many organizations provide estimates on the size of the industry, they are often not comparable to widely published industry data and statistics. This report examines the additive manufacturing industry in the U.S. and develops industry data that is comparable to that published by the U.S. Census Bureau. Additionally, it examines the adoption and diffusion of additive manufacturing technologies.

1.3. Scope and Approach

This report focuses on U.S. additive manufacturing; however, there is limited data on the nation's activities in this area. Wohlers[4] estimates that, globally, $1.714 billion in revenue was generated in the primary additive manufacturing market in 2011. This includes $834.0 million for additive manufacturing systems and materials; $642.6 million from the sale of parts produced from additive manufacturing systems; and $236.9 million for maintenance contracts, training, seminars, conferences, expositions, advertising, publications, contract research, and consulting. This report will focus on using these estimates combined with other figures to generate industry data on additive manufacturing that is comparable to industry data published by the U.S. Census Bureau. Data from the Annual Survey of Manufactures and methods developed by Thomas[5] are used in the development of industry data. The report also examines the adoption and diffusion of additive manufacturing by examining costs and unit sales.

There are variations between different types of additive manufacturing processes. These include photopolymer-based systems, powder-based systems, molten material systems, and solid sheet systems.[6] This report does not delve into the economic implications for each system. Rather it approaches additive manufacturing as a whole. Examining these system-related details would require additional research.

2. The U.S. Manufacturing Industry

Over time manufacturing processes have changed dramatically. Robotic arms and other machinery have radically changed the manufacturing environment. For instance, just a few decades ago a company such as Standard Motor Products, which produces replacement parts for car engines, had a number of employees who were illiterate. Today, many of the employees at Standard Motor Products not only need to be able to read, they need to know the computer language of the machinery producing the parts.[7,8] The increase in productivity that is often the result of these changes means fewer employees are needed to make the same products, possibly resulting in lower employment levels in manufacturing. And, while American manufacturing efficiency is improving, other nations have been developing and improving their own manufacturing industries. Emerging economies such as China have gone from producing some manufactured goods to producing a significant amount of goods. Understanding the current state and recent trends of the U.S. manufacturing industry in light of these issues is difficult. Tassey's "Rationales and Mechanisms for Revitalizing U.S. Manufacturing R&D Strategies"[9] and the commentaries that follow it, illustrate that determining the current and future state of U.S. manufacturing is controversial. Some experts have stated that U.S. multinationals have "abandoned" the U.S. and their global expansion "tends to 'hollow out'" U.S. operations while exporting jobs abroad. Others counter that operations and investment of U.S. multinationals are highly concentrated in the U.S. and maintain a large presence while increasing overseas activities.[10,11,12]

National economies are often compared to companies competing for market share. This is a common analogy made when discussing the U.S. manufacturing industry; unfortunately, this comparison can be rather misleading.[13,14,15,16,17] A national economy is the primary supplier of goods and services to its labor force, while a single company, generally, is not the primary supplier of goods and services to its employees. Additionally, a national economy provides the income for the majority of the nation's consumers, while a business, generally, does not provide the income for the majority of its customers. Moreover, a national economy represents a system of exchange in which a company operates as one entity of that system. Companies can go out of business while nations do not. Domestic demand for goods and services constitutes a great proportion of the demand for a nation's domestically-produced products, where the demand for goods and services from a company is primarily external. In addition to these types of analogies, frequently, anecdotal observations are used to characterize the manufacturing industry;[18] however, the insight from these types of observations is somewhat limited, as the manufacturing industry includes hundreds of thousands of establishments with millions of employees making trillions of dollars worth of goods. Anecdotal observations provide a limited narrow scope of the industry that does not necessarily reflect or apply to the industry as a whole.

The primary goal of devoting resources toward manufacturing activities is to receive a form of benefit for oneself and/or for society as a whole. This is true for all industry stakeholders. Investments are often assessed by the resources devoted to the investment and the resources that are yielded from the investment. The return is then compared to the return on other, similar, ventures. Also considered is the extent or size of one's investment. This is the approach that is taken in the following section to examine the manufacturing industry. Specifically, it examines the U.S. manufacturing industry from the stakeholder's return on investment and compares it internationally. This approach provides a systematic examination of the primary goal of devoting resources to manufacturing and sets it in the context of international performance.

2.1. The Current State of the Industry

According to 2010 data from the UN Statistics Division, the U.S. is the second largest manufacturing nation in the world, with China producing just slightly more than the U.S. as seen in Figure 2.1. This figure contains the ten largest manufacturing nations and illustrates the magnitude and significance of the U.S. manufacturing industry to the global and domestic economy. As seen in the pie charts, the U.S. produced 28% of the world's goods in 1985. This value declined to 18% in 2010. Although significant, it is important to note that in order for underdeveloped countries to become developed countries, their production and income will need to approach that of the developed world. This, inevitably, results in a decline in the proportion or market share that each developed country represents. In per capita terms, the U.S. is the fifteenth largest producer and far exceeds China (see Figure 2.2). However, the U.S. compound annual growth rate between 1985 and 2010 is 1.1%, putting it well below the 25th percentile of 181 nations as seen in Figure 2.3.

Using input-output analysis, the direct and indirect effects of U.S. manufacturing as a percent of output ranks 38th out of 45 countries; however, it is important to note that this value tends to decrease as nations increase their per capita gross domestic product (GDP). This does not suggest that manufacturing is less important to wealthy nations. While these effects decrease as a percent of output they increase on a per capita basis. Thus, high income nations tend to also have high levels of per capita manufacturing. The correlation coefficient between per capita GDP and manufacturing effects as a share of output is 0.846, suggesting a significant connection.

With the primary goal of devoting resources toward manufacturing activities being to gain a form of benefit for oneself and/or for society as a whole, the best variable to compare the return on investment to owners and financiers is net income per expenditure dollar; however, the primary variable available to examine and compare the returns for owners and financiers internationally is gross operating surplus per dollar of expenditure. Gross operating surplus is gross output less a subset of costs (i.e., intermediate expenditures, compensation, and taxes less subsidies), but does not take into account the depreciation of capital; therefore, it does not fully represent a return on investment. However, it is the best variable available. Employees exchange their time for compensation or income and consumers exchange the purchase price for the utility gained from the product purchased. Unfortunately, data is not readily available to examine the utility of consumers.

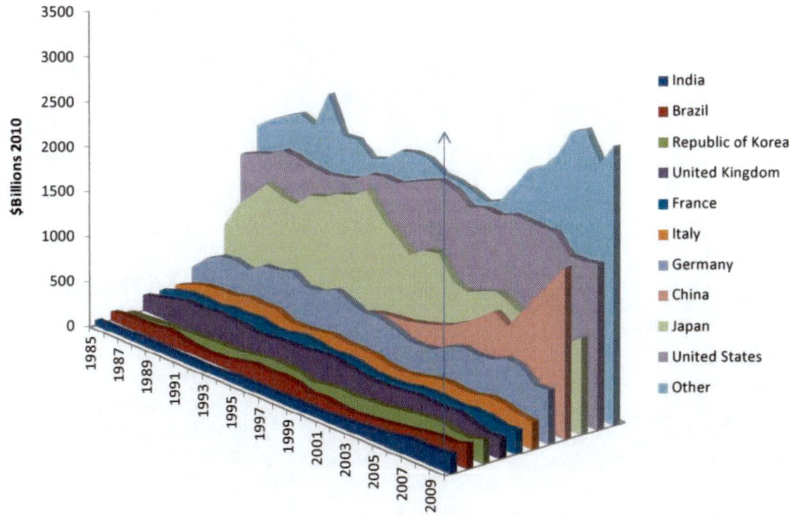

Figure 2.1. UNSD Manufacturing Value Added, Top Ten Producers.

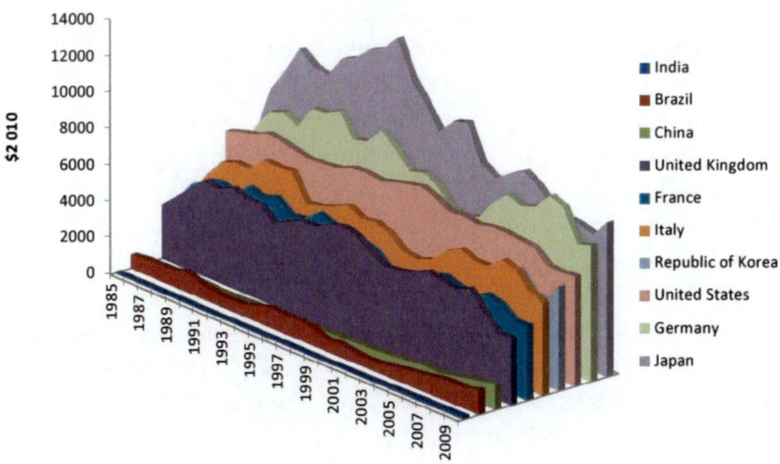

Figure 2.2. UNSD Manufacturing Value Added Per Capita, Top Ten Producers.

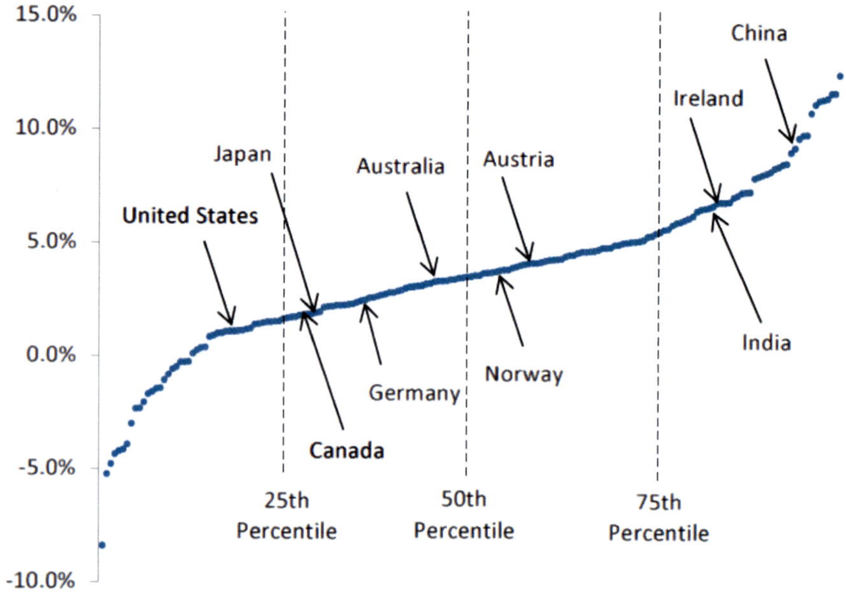

Figure 2.3. Manufacturing Value Added Compound Annual Growth, 1985-2010 (UNSD).

Among those countries for which data is available in the Organization for Economic Cooperation and Development's Structural Analysis (OECD STAN) database, Finland and Austria were the only countries to exceed the U.S. in gross operating surplus per expenditure dollar, compensation per hour, and manufacturing valued added per capita (Figure 2.4). Norway, Sweden, Germany, and Denmark have a higher per hour compensation and manufacturing value added per capita than the US, but have a significantly lower gross operating surplus per expenditure dollar. The U.S. manufacturing industry as a whole is just above the 62nd percentile for gross operating surplus per dollar of expenditure, with 14 out of 40 countries having a higher value. Compensation is ranked 9th among 20 countries for which data is available, putting the U.S. at the 55th percentile. The Netherlands, Norway, Sweden, Germany, Denmark, and France have higher levels of per hour compensation. For every dollar of manufacturing value added, there is an estimated 49.5 cents of value added from suppliers of goods and services. The gross operating surplus per expenditure dollar for suppliers was \$0.304 for the US, putting it at the 13th percentile. Indonesia had the highest level followed by Turkey, Greece, and Mexico. Compound annual growth in manufacturing between

1985 and 2010 is 1.1% putting it well below the 25th percentile of 181 nations; however, the U.S. continues to be the second largest manufacturing nation in the world, with China producing just slightly more than the US. In per capita terms, the U.S. is the fifteenth largest producer and far exceeds China. Its direct and indirect effects account for 28% of U.S. output.[19]

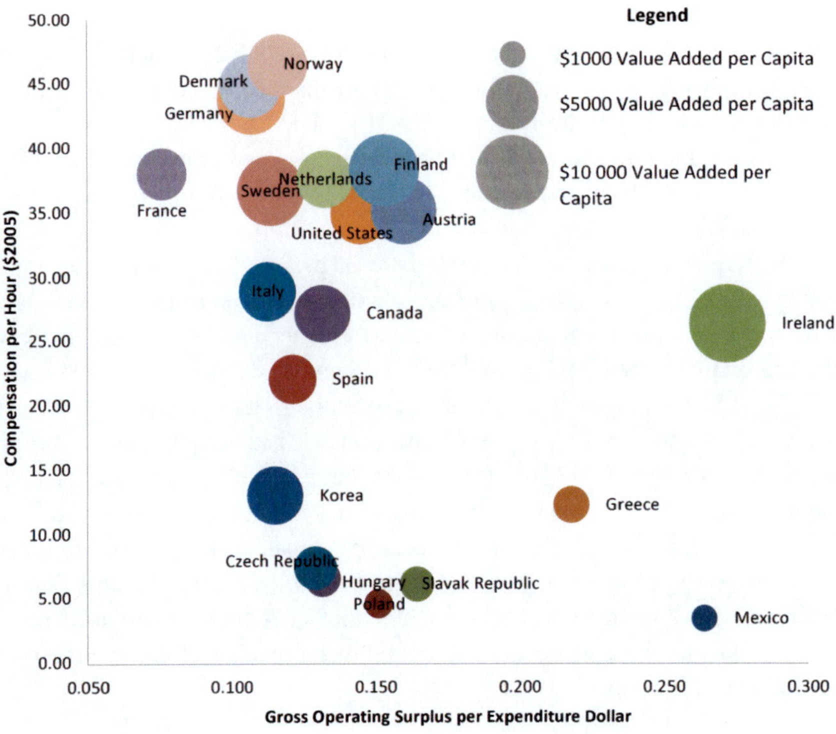

Figure 2.4. Manufacturing Value Added per Capita, Gross Operating Surplus per Expenditure Dollar, and Compensation per Hour, OECD STAN Data.

2.2. Science and Technology Innovation

According to the 2012 OECD Science and Technology Outlook, the U.S. is in a lead position for cutting-edge innovation. It maintains excellent higher education and leads the OECD in shares of gross domestic expenditure on research and development (R&D) (41%), triadic patent families[20] (29%), and scientific publications (31%). Large domestic firms contribute to an R&D intensive business sector amounting to 70% of total gross domestic expenditure on research and development. Small and medium enterprises account for 17% of business enterprise expenditures on research and development. Approximately 50% of research and development performers are in high-technology manufacturing.[21]

According to adjusted OECD STAN data, the U.S. has the largest research and development expenditure for total manufacturing among those countries for which data is available. In per capita terms, Germany spends nearly as much as the U.S. in research and development for all manufacturing, while Japan exceeds the U.S. expenditure by more than 30%. Among all OECD countries for which data are available, the U.S. ranks above the 95th percentile for total manufacturing research and development expenditures between 2001 and 2008. From 2001 through 2007, it was above the 90th percentile for all subsectors of manufacturing.[22]

OECD patent data includes the number of patents filed by the inventor's country of residence for 48 countries, including China and India as well as a world estimate. Patents reflect inventive performance and, therefore, are a key measure of innovation. According to OECD patent data, between 1999 and 2007 the U.S. has ranked above the 90th percentile in terms of total number of patents and above the 80th percentile in terms of patents per capita. During that same period, U.S. patents represented between 30% and 41% of total patents worldwide. This data is consistent with a patent analysis conducted by Thomson Reuters, which suggested that approximately 40% of the top 100 global innovator companies are located in the United States.[23] According to the OECD data, Japan is the only country that occasionally produced more patents than the U.S., while Luxembourg, Switzerland, and Japan produced more patents per capita in 2007.[24]

2.3. Additive Manufacturing

There is a general concern that the U.S. manufacturing industry has lost competitiveness with other nations; however, it still maintains a prominent position, as seen in the previous sections. The industry is the second largest in the world, but its growth is below the 25th percentile, placing it under that of Japan, Canada, Germany, and Australia among others. If the current trends in growth continue, by some measures, the U.S. manufacturing industry might be surpassed by other nations. According to the World Economic Forum's Global Competitiveness Index, the U.S. ranked 4th in global competitiveness in 2010-2011, 5th in 2011-2012, and 7th in 2012-2013, setting a downward trend.

Another concern is its rank in innovation which in 2012-2013 is 6th, down from 4th in 2010-2011.

Additive manufacturing may provide an important opportunity for advancing U.S. manufacturing while maintaining and advancing U.S. innovation. The U.S. is currently a major user of additive manufacturing technology and the primary producer of additive manufacturing systems. One of the major benefits of this technology is in the area of product design. It allows the production of nearly any complexity of geometry without the need for tooling. Additionally, the complexity does not impact the cost in the same way that it does for conventional manufacturing.[25] This technology eliminates many of the restrictions of 'Design for Manufacture and Assembly,' opening a new realm of possibilities for new customized products at an affordable price point.[26,27] To some degree, the success of this technology will rely on taking advantage of this benefit. With the U.S. being among the lead innovators and being the primary user of additive manufacturing, this technology may have the potential to significantly impact U.S. competitiveness.

Taking advantage of the opportunities that additive manufacturing offers may prove to be difficult. Designers and manufacturers have established practices and approaches to developing new products. Additive manufacturing presents new possibilities and, to some extent, requires new approaches. Changing the current practices in order to take advantage of new opportunities may be difficult. One such challenge is related to the customization of products to customer needs, which often requires a significant amount of input from the customer. Capturing this information could pose a new challenge to some manufacturers. Although the utility of consumers and end users is difficult to measure, these stakeholders will potentially be a major benefactor of additive manufacturing, as this technology enables rapid design-to-product transformation that enables new products to rapidly come to market.

Unfortunately, the available data does not allow an examination of the return on investment for stakeholders in additive manufacturing at this point in time. Section 4 discusses and estimates values for costs and profit; however, these are only reasonable approximations based on a combination of data sources. A comparison of return on investment using this data would not represent the true state of U.S. additive manufacturing.

Additive manufacturing may make the U.S. a more competitive place for manufacturing resulting in more goods being produced in the U.S.; however, it is important to note that productivity is a contributor to the reduction of manufacturing employment.[28] Even if additive manufacturing results in a significant increase in productivity that attracts jobs from overseas, it may not result in a net increase in manufacturing employment; however, it is possible that additive manufacturing may facilitate a net increase in employment through new products or other means.

3. Additive Manufacturing Stakeholders

This section identifies stakeholders and costs related to additive manufacturing. These items are relevant to understanding the adoption and diffusion of this technology. Individual manufacturing stakeholders are affected by the industry in different ways. Therefore, it is useful to identify individual stakeholders and classify them into stakeholder groups. This classification can then be used to identify the primary investment each stakeholder has in the manufacturing industry along with their expected return. Stakeholders evaluate benefits and costs of manufacturing industry investments purely from their "stakeholder" point of view; therefore, it is important to identify each stakeholder's investment and expected return. These perspectives can provide some guidance to the adoption of additive manufacturing.

There are a number of stakeholders for the additive manufacturing industry. The most direct and obvious ones are the owners and employees of manufacturing companies; these are the individuals directly responsible for production. As seen in the manufacturing supply chain in Figure 3.1, there are many suppliers of goods and services that also have a stake in the industry; these include resellers, providers of transportation and warehousing, raw material suppliers, suppliers of intermediate goods, and suppliers of professional services. The items in the figure colored in blue represent suppliers of services, computer hardware, software, and other costs. Tan represents refuse removal, intermediate goods, and recycling, while orange represents machinery, structures, and compensation, with red being the repair of the machinery and structures. Green represents the suppliers of materials. These items all feed into the design and production of manufactured goods that are inventoried and/or shipped. The depreciation of capital and net income are also included in the figure, which affect the market value of shipments. In addition to the stakeholders in the figure, there are also public vested interests, the end users, and financial service providers.

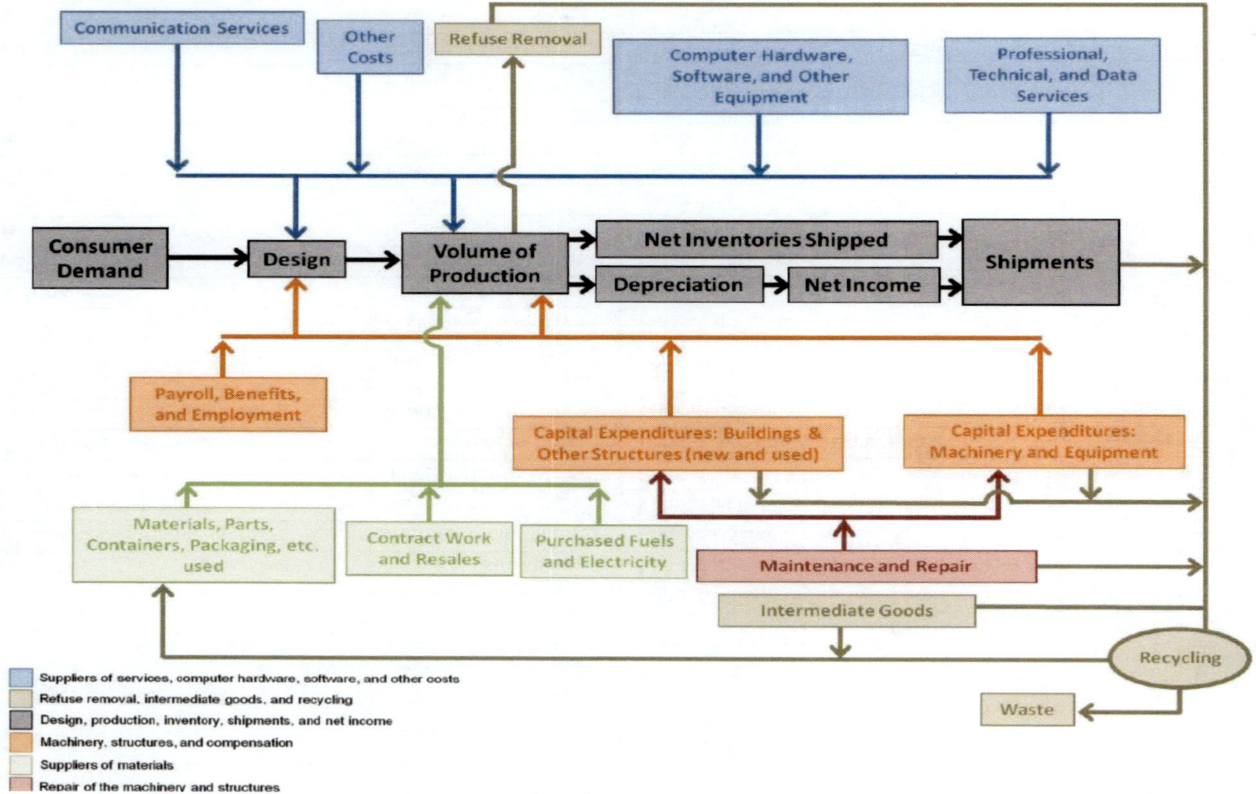

Figure 3.1. Manufacturing Supply Chain.

Table 3.1. Stakeholders

Stakeholders	Affiliation	Primary Investment	Expected Return
Owners	Private Producers	Land, Capital Goods, and Financial Capital	Profit From Sales
Employees (manufacturing industry and suppliers)	Laborers	Labor	Income
Resellers	Private Distributer	Land, Capital Goods, and Labor	Profit From Markup
Retailers	Private Distributer	Land, Capital Goods, and Labor	Profit From Markup
Wholesalers	Private Distributer	Land, Capital Goods, and Labor	Profit From Markup
Standards and Codes Organizations	Public/Private Interest	Labor and Intellectual Property	Economic Success
Transportation and Warehousing	Support Service	Land, Capital Goods, and Labor	Profit From Fees
Air Transportation Providers	Transportation	Land, Capital Goods, and Labor	Profit From Fees
Ground Transportation Providers	Transportation	Land, Capital Goods, and Labor	Profit From Fees
Warehousing and Storage Providers	Storage Facility	Land and Capital Goods	Profit From Fees
Professional Societies	Public/Private Support Services	Labor and Intellectual Property	Economic Success and Profit from Fees
Finance Services	Insurance and Finance	Financial Capital	Profit From Fees
Insurance Providers	Insurance	Financial Capital	Profit From Fees
Health and Medical Insurance Providers	Insurance	Financial Capital	Profit From Fees
Financiers	Financier	Financial Capital	Capital Gains
Public Vested Interests	Public	Labor and Financial Capital	Economic Success
Policy Makers	Public	Labor and Financial Capital	Economic Success
Tax Payers	Public	Financial Capital	Economic Success
Industry Suppliers	Public/Private Suppliers	Land, Capital Goods, and Labor	Profit
Mining Material Suppliers	Private Suppliers	Land, Capital Goods, and Labor	Profit From Sales

Stakeholders	Affiliation	Primary Investment	Expected Return
Agriculture Product Suppliers	Private Suppliers	Land, Capital Goods, and Labor	Profit From Sales
Electric Utility Suppliers	Private Suppliers	Land, Capital Goods, and Labor	Profit From Sales
Water Utility Suppliers	Public/Private Suppliers	Land, Capital Goods, and Labor	Profit From Sales
Natural Gas Suppliers	Private Suppliers	Land, Capital Goods, and Labor	Profit From Sales
Facility Construction Providers	Private Suppliers	Land, Capital Goods, and Labor	Profit From Sales
Maintenance and Repair Providers	Private Suppliers	Land, Capital Goods, and Labor	Profit From Sales
Communication Services Providers	Private Support Services	Land, Capital Goods, and Labor	Profit From Fees
Other Fuel Suppliers	Private Suppliers	Land, Capital Goods, and Labor	Profit From Sales
Refuse Removal Service Providers	Private Support Services	Land, Capital Goods, and Labor	Profit From Fees
Professional Services	Public/Private Support Services	Land, Capital Goods, Labor, and Intellectual Property	Profit From Fees
Legal Service Providers	Public/Private Support Services	Labor	Profit From Fees
Information Service Providers	Private Support Services	Land, Capital Goods, and Labor	Profit From Fees
Research Organizations	Public/Private Suppliers	Labor and Intellectual Property	Profit From Fees
Accounting Service Providers	Private Support Services	Labor	Profit From Fees
Engineering Service Providers	Private Support Services	Labor and Intellectual Property	Profit From Fees
Computer Service Providers	Private Support Services	Labor	Profit From Fees

Table 3.1. (Continued)

Stakeholders	Affiliation	Primary Investment	Expected Return
Scientific and Technical Service Providers	Private Support Services	Labor and Intellectual Property	Profit From Fees
Advertisers	Private Support Services	Labor and Intellectual Property	Profit From Fees
Other Professional Services	Private Support Services	Labor and Intellectual Property	Profit From Fees
Consumers/End User	End User	Product Purchasing Price	Final Product Utilization

As seen in Table 3.1, stakeholders may have a direct investment in manufacturing, such as industry owners and employees, or an indirect investment through supply chains or industry outputs. Each stakeholder is associated with a primary form of investment. For example, employees invest their labor, while owners invest land and capital. Owners often have labor and/or intellectual property invested as well; however, their primary investment is in the form of land and capital as seen in Table 3.1. Each stakeholder has invested these items with the expectation of receiving compensation or a return on investment. Employees, for instance, expect to be compensated for their labor and owners expect to receive a profit. There are six different categories of assets used in Table 3.1 that can be vested into the industry: financial capital, capital goods, land, labor, intellectual property, and the end users purchasing price. A successful industry might be considered one that has a suitable magnitude of production that results in competitive net benefits for its stakeholders. The expected returns from the industry include profits from sales, markup, or fees; income; industry success; capital gains; and utility from the final use of the product.

Summary of Primary Investments

Land: Naturally-occurring goods such as water, air, soil, mineral, and flora used in the production of products (i.e., the totality of goods or services that a company makes available).

Labor: Human effort used in production, which includes technical and marketing expertise.

Capital Goods: Human made goods used in the production of products.

Financial Capital: Funds provided by investors to purchase capital goods for production of products.

Intellectual Property: Ideas, trademarks, copyrights, trade secrets, and patents used to produce products

Purchasing Price: Market value of products sold

Summary of Expected Returns

Profit from sales: The financial benefit realized when revenues exceed costs and taxes for a product.

Capital Gains: An increase in the value of a capital asset Income: Compensation for an individual's service or labor

Profit from Markup: The difference between the cost of a product and its selling price.

Economic Success: A constant and suitable magnitude of production resulting in competitive benefits (profits, capital gains, income, and product utilization) for an industry's stakeholders.

Profit from Fees: The financial benefit realized when revenues exceed costs and taxes for a service.

Table 3.2 provides a list of stakeholders and the potential impact additive manufacturing might have on them. The adoption of additive manufacturing is likely to have a significant impact on the consumer/end user, as this technology improves new products and facilitates the rapid production of new products. These individuals will be the primary beneficiaries of customized complex products that meet their individualized needs.

Financiers, employers, and suppliers will benefit from the profit of new product sales; however, some of the new products will be replacing previously produced products and the source of revenue might just shift from one product to another. Additionally, any increased profit commanded from these products will be partially reduced through competition as more companies enter the market. The benefit of new customized complex products, however, will continue to benefit end users. It is possible that some of the largest benefits of additive manufacturing will be realized outside of the manufacturing industry.

Table 3.2 also provides a list of costs to stakeholders, as the development and use of additive manufacturing technology has some costs associated with it. The owners invest in the research and development of this technology and also must purchase new machinery to replace traditional manufacturing machinery. Resellers may have to bear the burden of gathering information from customers for customized products. Some of these costs may be passed on to the consumers/end users through the purchase price.

Table 3.2. Stakeholder Benefits for Adopting Additive Manufacturing

Stakeholders	Primary Benefits of the Adoption of Additive Manufacturing	Primary Costs to the Adoption of Additive Manufacturing
Owners	New product sales, increased efficiency and productivity	Cost of research and development, new machinery costs
Employees (manufacturing industry and suppliers) Reshoring of jobs, increase in income	Labor, possible decrease in employment	
Resellers	New product sales	Cost of gathering consumer data for customized products
Retailers	New product sales	Cost of gathering consumer data for customized products
Wholesalers	New product sales	Cost of gathering consumer data for customized products
Standards and Codes Organizations	Economic success	Cost of research and development
Transportation and Warehousing	Increased demand, reduced vehicle weight	Cost of new products
Air Transportation Providers	Increased demand, reduced vehicle weight	Cost of new products
Ground Transportation Providers	Increased demand, reduced vehicle weight	Cost of new products
Warehousing and Storage Providers	Increased demand	Decreased demand
Professional Societies	Economic success	Cost of research and development
Finance Services	Profit, product reliability and reduced claims, increased demand	Initial investment
Insurance Providers	Product reliability and reduced claims	Minimal cost
Health and Medical Insurance Providers	Increased demand for services	Cost of research and development

Table 3.2. (Continued)

Stakeholders	Primary Benefits of the Adoption of Additive Manufacturing	Primary Costs to the Adoption of Additive Manufacturing
Financiers	Profit from fees and capital gains	Initial investment
Public Vested Interests	Economic Success, increased standard of living	Cost of research and development
Policy Makers	Economic Success, increased standard of living	Cost of research and development
Tax Payers	Economic Success, increased standard of living	Cost of research and development
Industry Suppliers	Increased demand	Cost of meeting increased demand
Mining Material Suppliers	Increased demand	Cost of meeting increased demand
Agriculture Product Suppliers	Increased demand	Cost of meeting increased demand
Electric Utility Suppliers	Increased demand	Cost of meeting increased demand
Water Utility Suppliers	Increased demand	Cost of meeting increased demand
Natural Gas Suppliers	Increased demand	Cost of meeting increased demand
Facility Construction Providers	Increased demand, new construction materials	Cost of meeting increased demand
Maintenance and Repair Providers	Possible increased demand	Cost of meeting increased demand, possible decrease in demand
Communication Services Providers	Increased demand	Cost of meeting increased demand
Other Fuel Suppliers	Increased demand	Cost of meeting increased demand
Refuse Removal Service Providers	Reduced vehicle weight	Cost of meeting increased demand, possible decrease in demand
Professional Services	Increased demand	Cost of meeting increased demand
Legal Service Providers	Increased demand	Cost of meeting increased demand

Stakeholders	Primary Benefits of the Adoption of Additive Manufacturing	Primary Costs to the Adoption of Additive Manufacturing
Information Service Providers	Increased demand	Cost of meeting increased demand
Research Organizations	Increased demand	Cost of meeting increased demand
Accounting Service Providers	Increased demand	Cost of meeting increased demand
Engineering Service Providers	Increased demand	Cost of meeting increased demand
Computer Service Providers	Increased demand	Cost of meeting increased demand
Scientific and Technical Service Providers	Increased demand	Cost of meeting increased demand
Advertisers	Increased demand	Cost of meeting increased demand
Other Professional Services	Increased demand	Cost of meeting increased demand
Consumers/End User	New product utilization, cost reduction, increased efficiency	Increased purchase price

4. Industry Use of Additive Manufacturing

Value added is the best measure available for comparing the relative economic importance of manufacturing among various industries, as it avoids the duplication caused from the use of products of some establishments as materials in others. The Annual Survey of Manufactures, one of the datasets used in this report, calculates value added as the value of shipments less the cost of materials, supplies, containers, fuel, purchased electricity, and contract work (i.e., shipments less the suppliers of materials colored green in Figure 4.1). It is adjusted by the addition of value added by merchandising operations plus the net change in finished goods and work-in-process goods. It is important to note that this calculation of value added varies from that of other organizations. The U.S. Bureau of Economic Analysis (BEA), for example, calculates value added as "gross output (sales or receipts and other operating income, plus inventory change) less intermediate inputs (consumption of goods and services purchased from other industries or imported)."[29] The primary difference is that the Annual Survey of Manufacture's calculation of value added includes purchases from other industries such as mining and construction while BEA and other organizations do not include it (i.e., BEA calculates it as shipments less all costs colored blue, tan, orange, red, and green in Figure 4.1). Since this report uses data from the Annual Survey of Manufactures, it will maintain their method of calculating value added.

Although value added is discussed, most of the figures in this report are in terms of shipments, which is analogous to revenue. This value is used because the data collected on additive manufacturing is in terms of revenue; thus, in order to discuss value added, additional assumptions must be made, which introduces additional imprecision.

4.1. Products of Additive Manufacturing

Globally, an estimated $642.6 million in revenue was collected for additive manufactured goods[30] with the U.S. accounting for an estimated $246.1 million or 38.3% of global production in 2011. [31] As seen in Table 4.1, these products are categorized as being in the following sectors: motor vehicles; aerospace; industrial/business machines; medical/dental; government/military; architectural; and consumer products/electronics, academic institutions, and other. The consensus among well-respected industry experts is that the penetration of the additive manufacturing market is 8%;[32]

however, as seen in Table 4.1, goods produced using additive manufacturing methods represent between 0.01% and 0.05% of their relevant industry subsectors. Thus, additive manufacturing has sufficient room to grow.

Figure 4.1 provides an estimated supply chain for products of additive manufacturing using the methods documented in NIST Special Publication 1142 combined with some additional assumptions.[33] The estimation method used provides rough estimates; thus, some caution should be used. Additional precision would require further data collection. The items in the figure colored in blue represent suppliers of services, computer hardware, software, and other costs. Tan represents refuse removal, intermediate goods, and recycling, while orange represents machinery, structures, and compensation, with red being the repair of the machinery and structures. Green represents the suppliers of materials. These items all feed into the design and production of manufactured goods that are inventoried and/or shipped. The depreciation of capital and net income are also included in the figure, which affect the market value of shipments. The net income per expenditure dollar (i.e., return on investment) is approximately 0.205; however, this may have significant variation. The total number of employees estimated in U.S. additive manufacturing products is estimated at 658. The following sections discuss the various categories of manufacturing that use this technology.

Motor Vehicles

Shipments for the U.S. automotive industry (NAICS 3361, 3362, and 3363) was estimated at $445 billion in 2011. Approximately 19.5% of additive manufacturing is within the automotive industry, with the U.S. share being estimated as $48.0 million or 0.01% of the U.S. automotive industry. The industry frequently uses additive manufacturing technologies for rapid prototyping. It is also commonly used for complex, high-value, or custom parts for antique cars. Motorsports such as NASCAR and Formula 1 have also been a field for the application of this technology, which have some crossover with the aerospace industry. Both sectors have high demand for performance and weight reduction.

Examples of motor vehicle applications include the following:

- Intake valves, engine bay parts, gear boxes, and engine components
- Air inlet, engine control unit and lower fairing baffle
- Testing of parts
- Motorcycle engines

Table 4.1. Additive Manufacturing Shipments

Category	Relevant NAICS Codes	Percent of Total AM Made Products	Shipments of US Made AM Products ($millions, 2011)*	Total Shipments ($millions, 2011)	AM Share of Industry Shipments
Motor vehicles	NAICS 3361, 3362, 3363	19.5%	48.0	445 289.4	0.01%
Aerospace	NAICS 336411, 336412, 336413	12.1%	29.8	157 700.7	0.02%
Industrial/business machines	NAICS 333	10.8%	26.6	365 734.8	0.01%
Medical/dental	NAICS 3391	15.1%	37.2	89 519.5	0.04%
Government/military	NAICS 336414, 336415, 336419, 336992	6.0%	14.8	32 784.4	0.05%
Architectural	NAICS 3323	3.0%	7.4	72 186.9	0.01%
Consumer products/electronics, academic institutions, and other	All other within NAICS 332 through 339	33.6%	82.7	895 709.8	0.01%
TOTAL	NAICS 332 through 339	100.0%	246.1	2 058 925.5	0.01%

* These values are calculated assuming that the percent of total additive manufacturing made products for each industry is the same for the U.S. as it is globally. It is also assumed that the U.S. share of AM systems sold is equal to the share of revenue for AM products

Note: Numbers may not add up to total due to rounding.

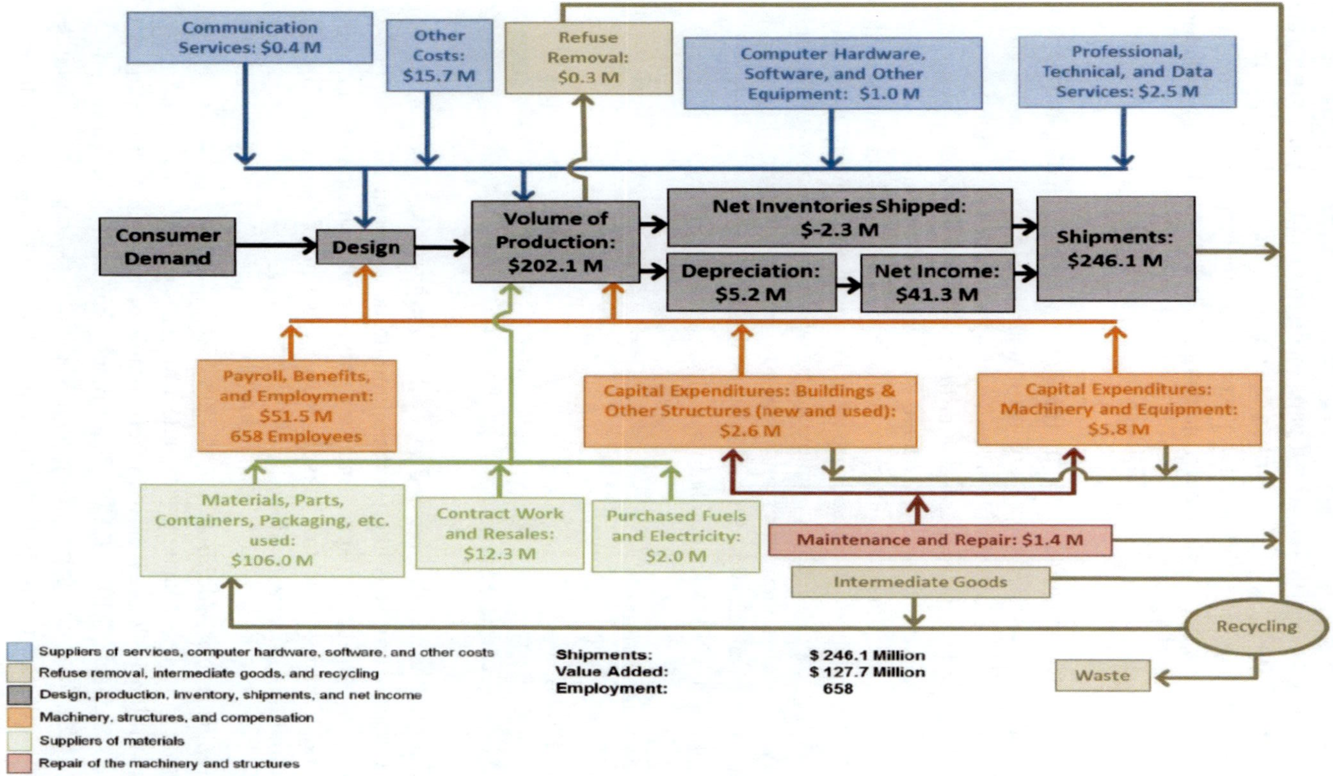

Figure 4.1. Supply Chain for Additive Manufacturing Products, 2011.

The restricted construction size of parts made from additive manufacturing has been a limiting factor for further adoption of this technology in the automotive industry. As the additive manufacturing industry develops the ability to produce larger components, the automotive industry is likely to adopt this technology more rapidly.[34,35,36]

Aerospace

Shipments for manufacturing in the U.S. aerospace industry (NAICS 336411, 336412, and 336413) were estimated at $157.7 billion in 2011. Approximately 12.1% of additive manufacturing is within this industry, with the U.S. share being estimated as $29.8 million or 0.02% of the U.S. aerospace industry. Aerospace includes a range of vehicles including airplanes, unmanned vehicles, transport vehicles, and space vehicles. This industry has significant potential for increased use of additive manufacturing as it often requires strong geometrically complex parts, which must be especially light weight. Additionally, these parts are, typically, produced in small quantities, making them a likely candidate for additive manufacturing.

Examples of aerospace applications include the following:

- Structural parts
- Thrust reverser doors
- Landing gears
- Gimbal eye
- Fuel injection nozzles

Similar to the automotive industry, the restricted construction size of additive manufacturing has likely been a limiting factor for further adoption of this technology in the aerospace industry. Additionally, materials, accuracy, surface finish, and certification standards have also played a role in limiting further adoption of this technology.[37,38,39,40]

Industrial/Business Machines

Shipments in U.S. machinery manufacturing (NAICS 333) were estimated at $365.7 billion in 2011. Approximately 10.8% of additive manufacturing is within this industry, with the U.S. share being estimated at $26.6 million or 0.01% of U.S. machinery manufacturing. Machinery manufacturing includes the creation of end products that appl-y mechanical force to perform work. Additive manufacturing technology has been used in the development and production of parts for these machines. For example, a new drag chain link was developed and produced for the mining industry using additive manufacturing.

Medical/Dental

Shipments for U.S. manufacturing of medical and dental products (NAICS 3391) amounted to $89.5 billion in 2011. Approximately 15.1% of additive manufacturing is within this industry, with the U.S. share being estimated at $37.2 million or 0.04% of medical/dental manufacturing. The need for custom-made products in the medical and dental industry creates a demand for products made using additive manufacturing methods. Items produced include custom implants, prosthetics, surgical tools, hearing aids, and drug delivery devices among other items. Emerging research and development has resulted in biomanufacturing, where the construction of tissue from living cells is used to “print” organs. Although this field is not fully developed, it is a promising area for applying additive manufacturing technology.

Government/Military

Shipments for U.S. manufacturing of products for the government and military (NAICS 336414, 336415, 336419, 336992) amounted to $32.8 billion in 2011. Approximately 6.0% of additive manufacturing is within this industry, with the U.S. share being estimated at $14.8 million or 0.05% of government/military manufacturing. The U.S. military has shown interest in advancing research and procurement of additive manufacturing for a number of components. The U.S. Air Force, for example, is conducting research on the use of additive manufacturing for metal parts, heat exchangers, and plastic resins for remotely piloted vehicles. The U.S. Navy is also investigating the use of this technology.[41]

Architecture

Shipments for U.S. manufacturing of products for architecture (NAICS 3323) amounted to $72.1 billion in 2011. Approximately 3.0% of additive manufacturing is within this industry, with the U.S. share being estimated at $7.4 million or 0.01% of architectural manufacturing. A major use of additive manufacturing for architecture is in the modeling of structures and designs. In the past, physical models were tediously built by hand. Additive manufacturing has revolutionized this process.

Consumer Products/Electronics, Academic Institutions, and Other

Shipments for U.S. manufacturing for consumer products/electronics, academic institutions, and other amounted to $895.7 billion in 2011. Approximately 33.6% of additive manufacturing is within this industry, with the U.S. share being estimated at $82.7 million or 0.01% of this category of manufacturing. It includes many items produced using additive manufacturing technology, including toys, figurines, furniture, office accessories, musical instruments, art, jewelry, museum displays, and fashion products among other items.

4.2. Additive Manufacturing Systems

Approximately 62.8% of all commercial/industrial units sold in 2011 were made by the top three producers of additive manufacturing systems: Stratasys, Z Corporation[42], and 3D Systems based out of the U.S. Approximately 64.4% of all systems were made by companies based in the U.S. The total global revenue from system sales was $502.5 million with U.S. revenue estimated at $323.6 million as seen in Figure 4.2.[43] The production of additive manufacturing systems or 3D printers can be categorized as being under NAICS 332: Industrial Machinery Manufacturing. Data from the Annual Survey of Manufactures for this sector was used to develop the estimates in Figure 4.2. The net income as a share of revenue (i.e., shipments) for Stratasys and 3D Systems, two of the three largest additive manufacturing system producers, was 0.144 and 0.178, while the estimate using data in Figure 4.2 is 0.152.

It is important to remember that additive manufacturing systems are already incorporated into the sales of products produced using this technology; thus, it would be unorthodox to add the value for additive manufactured products together with the value for the systems.

4.3. Additive Manufacturing Costs

Manufacturing processes and manufacturing parts are becoming more and more complex. Additive manufacturing both reduces and adds to the complexity of this process. As seen in Table 4.2, there are a number of pros and cons involved with additive manufacturing.

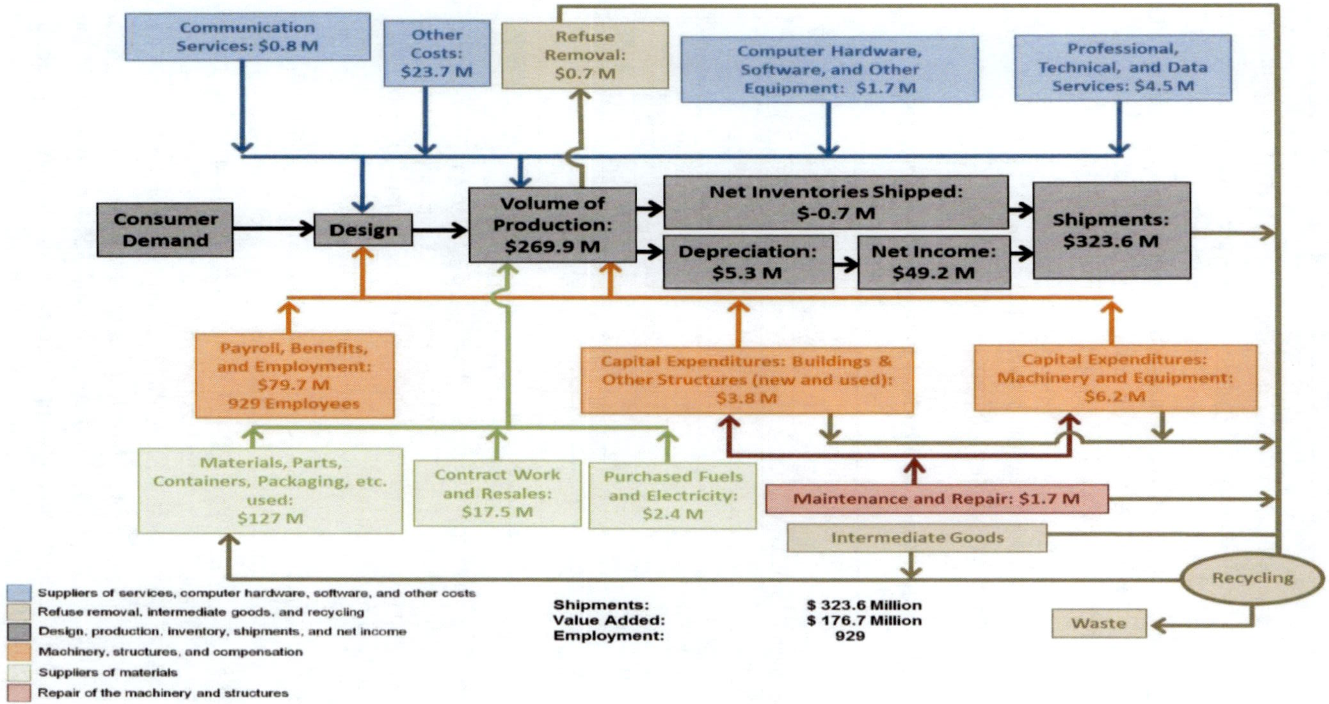

Figure 4.2. Supply Chain for Additive Manufacturing Systems, 2011.

For instance, there are fewer parts to manage, more flexibility in design, and products can be individualized; however, there are higher calibration requirements, needed quality improvements, and parts often require reworking. The benefits of additive manufacturing are not limited to the producer, however, as the end user also benefits from increased functionality, reduced lifecycle costs, and new product utilization. Aerospace parts, for instance, have shown a weight reduction potential of up to 70% of the original part[44] and a 1 kg reduction in weight saves an estimated $3000 of fuel annually[45], not to mention the reduction in emissions.

Table 4.2. Pros and Cons in Product Lifecycle Management

Pros	Cons
More flexible development	Software limitations
Freedom of design and construction	High machine and material costs
Integration of functions	High calibration effort
Less assembly	Deficient quality
Fewer production tools necessary	Parts often require reworking
Less spare parts in stock	Building time depends on part height
Less complexity (fewer parts to manage)	
Fewer tools needed	
Less time-to-market	
Rapid alterations	

Source: Lindemann C., U. Jahnke, M. Moi, and R. Koch. “Analyzing Product Lifecycle Costs for a Better Understanding of Cost Drivers in Additive Manufacturing.” Proceedings of the 2012 Solid Freeform Fabrication Symposium. <http://utwired.engr.utexas.edu/lff/symposium/proceedingsArchive/pubs/Manuscripts/2012/2012-12- Lindemann.pdf>

Costs have been identified as being a significant factor in whether producers adopt additive manufacturing technologies. Hopkinson estimates that machine costs range between 50% and 75% of total cost, materials range between 20% and 40%, and labor ranges between 5% and 30%.[46] The price for materials can vary somewhat. Stereolithography/epoxy-based resin is estimated at $175 per kilogram, selective laser sintering/nylon powder is $75, and fused deposition modeling/ABS filament is around $250. To put this in the perspective of conventional manufacturing, injection molding/ABS is about $1.80 and machining/1112 screw-machine steel is about $0.66.[47]

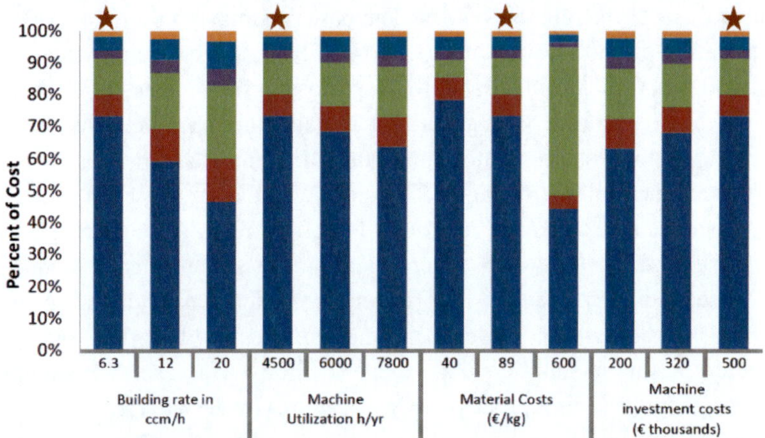

	6.3	12	20	4500	6000	7800	40	89	600	200	320	500
Building process fix	2%	2%	3%	2%	2%	2%	2%	2%	1%	2%	2%	2%
Preparation costs	4%	7%	9%	4%	5%	6%	5%	4%	3%	6%	5%	4%
Oven	3%	4%	5%	3%	3%	4%	3%	3%	2%	4%	3%	3%
Material costs	11%	17%	23%	11%	14%	16%	5%	11%	46%	16%	14%	11%
postprocessing costs	7%	10%	13%	7%	8%	9%	7%	7%	4%	9%	8%	7%
Machine costs	74%	59%	47%	74%	69%	64%	78%	74%	45%	63%	68%	74%

Source: Lindemann C., U. Jahnke, M. Moi, and R. Koch. "Analyzing Product Lifecycle Costs for a Better Understanding of Cost Drivers in Additive Manufacturing." Proceedings of the 2012 Solid Freeform Fabrication Symposium. <http://utwired.engr.utexas.edu/lff/symposium/proceedingsArchive/pubs/Manuscripts/2012/2012-12- Lindemann.pdf>

Note: The orange star indicates the base model.

Figure 4.3. Cost Distribution of Additive Manufacturing of Metal Parts by varying Factors.

Other research on metal parts confirms that machine and material costs are a major cost driver for this technology as seen in Figure 4.3, which presents data for a sample part made of stainless steel. For this example, four cost factors are varied and the production quantity is a little less than 200 for the base case. This analysis provides insight into identifying the largest costs of additive manufacturing. The first cost factor that is varied is the building rate, which is the speed at which the additive manufacturing system operates. In this example, it is measured in cubic centimeters per hour. The second factor that is varied is the machine utilization measured as the number of hours per year that the machine is operated. The third factor is the material cost and the last factor is the machine investment costs, which include items related to housing, using, and maintaining the additive manufacturing system. Among other things, this includes energy costs, machine purchase, and associated labor costs to operate the system. The base model has a build rate of 6.3 ccm, a utilization of 4500 h/yr, a material cost of 89 €, and a machine investment cost of 500 000 €. For comparison, the base case is shown four times in the figure, with each one shown with a star. On average, the machine costs accounted for 62.9% of the cost estimates in Figure 4.3 (note that the base case is only counted once in the average). This cost was the largest even when building rate was more than tripled and other factors were held constant. This cost was largest in all but one case, where material costs were increased to 600 €/kg. The second largest cost is the materials, which, on average, accounted for 18.0% of the costs; however, it is important to note that this cost is likely to decrease as more suppliers enter the field.[48] Post processing, preparation, oven heating, and building process fix were approximately 8.4%, 5.4%, 3.3%, and 1.9%, respectively.

Plastic parts likely have a slightly different cost structure. A case study of a fluorescent lamp holder provides some insight. This case study examined two Electro Optical Systems that use selective laser sintering: P390 and P730. The P390 was more cost effective for this particular case study. The cost per part for this item was examined and revealed that for the P390, 58.7% of the cost was machine cost, 9.9% was machine operator cost, 30.4% was material cost, and 1.0% was assembly.[49]

For manufacturers, the cost advantage of additive manufacturing may vary. Typically, it is believed that this technology is competitive for low volume production. This can be illustrated in another case study of a landing gear assembly for a 1:5 scale model of the P180 Avant II by Piaggio Aero Industries S.p.A. As seen in Table 4.3, the per assembly cost of producing the landing gear using traditional manufacturing methods, in this case high pressure die casting, was 21.29 € plus 21 000 € divided by the lot size. The cost of additive manufacturing was 526.31 € per assembly; thus, below a lot size of 41 additive manufacturing is more cost effective. Above a lot size of 41 it was not cost effective. These cost estimates also illustrate how additive manufacturing does not follow traditional economies of scale, where large production runs reduce the per item cost; thus, each assembly produced using additive manufacturing costs the same regardless of how many are produced. The cost effectiveness of using additive manufacturing relies on a number of factors, including the complexity of the part, amount of material, and the volume of production.

Table 4.3. High Pressure Die Cast Manufacturing Costs vs. Additive Manufacturing Costs (Selective Laser Sintering)

	Traditional Manufacturing (High Pressure Die Cast)	Additive Manufacturing (Selective Laser Sintering)
Material cost per part Mould cost per part	2.59 € 21 000 €/N	25.81 € -
Pre-processing cost per part	-	8.00 €
Processing cost per part	0.26 €	472.50 €
Post-processing cost per part	17.90 €	20.00 €
Linkages and assembly	0.54 €	-
TOTAL COST PER ASSEMBLY	21.29 €+21 000 €/N	526.31 €

Note: N is the lot size or the number of consecutive assemblies produced.

Source: Atzeni, Eleonora, Luca Iuliano, and Alessandro Salmi. (2011) "On the Competitiveness of Additive Manufacturing for the Production of Metal Parts." Proceedings of the 9th International Conference on Advanced Manufacturing Systems and Technology.

5. Adoption and Diffusion of Additive Manufacturing

5.1. The Diffusion Process

Disseminating a new idea or innovation so that it is widely adopted can be difficult, even if it has obvious advantages. A common challenge for many is how to speed up the rate of diffusion of an innovation. *Diffusion*, for the purpose of this report, is defined as, "the spread of an innovation throughout a social system," while *adoption* is defined as, "the acceptance and continued use of a product, service, or idea."[50] The diffusion of new technologies or innovations tends to follow certain trends and the process is studied in several disciplines: economics, communications, sociology, and marketing.

There is both a diffusion model and an adoption model. The diffusion model is illustrated by the logistic S-curve that evaluates the time it takes for an innovation to be diffused into an industry.[51] Diffusion increases at an increasing rate up to time T_1, and then at a decreasing rate thereafter (see Figure 5.1).

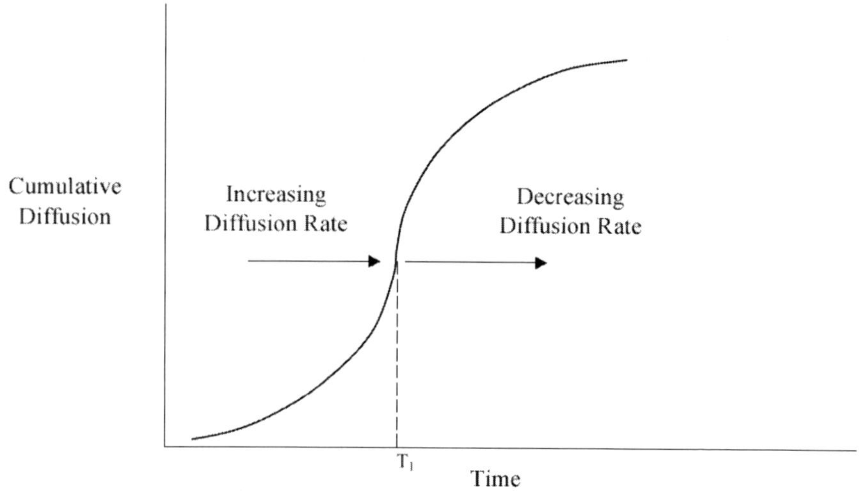

Modified from Rogers, E. M., (1995). *Diffusion of Innovations,* Fourth Edition, (New York: The Free Press, 1995), 258.

Figure 5.1. The Logistical S-Curve Model of Diffusion.

A simple logistic function may be defined by the following equation:

$$P(t) = \frac{1}{1 + e^{-t}}$$

Where P represents the population of adopters and t is time. The early growth is exponential and decays after 50% of adopters are reached and e is Euler's number, the base of the natural system of logarithms.[52]

In connection with the diffusion model, the adoption model focuses on the decision process of the individual or firm. This model is connected with Everett Rogers' theory[53] that the S-curve is normally distributed (see Figure 5.2). Most adopters act in the midrange of the adoption period timeline because of information diffusion. This is where the adoption rate is the highest. At the "early adopters" stage in Figure 5.2, relatively little is known about the new technology and the number of adopters is low. At the stage of the "majority of adopters," a significant amount of information has been diffused. By the "late adopters" stage, there is little information remaining to be diffused. Each individual's adoption of the technology is equivalent to a "learning trial" in the system. Over time, adopter distributions follow a bell shaped curve.

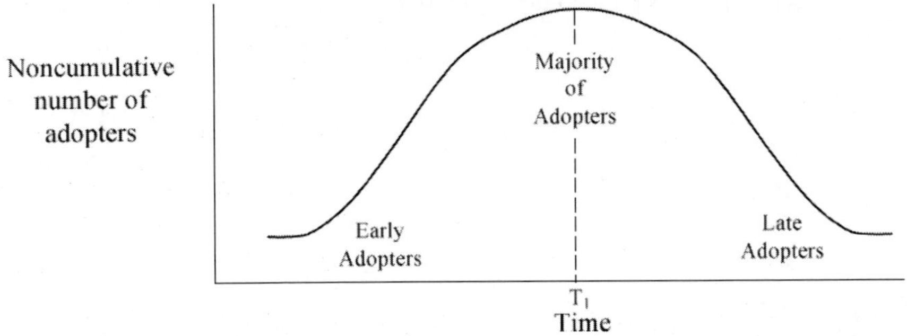

Modified from Rogers, E. M. (2003). *Diffusion of Innovations,* 5th Edition (New York: The Free Press, 2003), 111- 114.

Figure 5.2. Rogers' Model of Adoption (based on probability distribution).

Larsen stresses three explanatory innovation diffusion concepts: (1) cohesion, (2) structural equivalence, and (3) thresholds.[54] *Cohesion* asserts that diffusion takes place by face-to-face contact between stakeholders, who are described as sharing a high degree of *homophily*; that is to say, they have a tendency to listen to people similar to themselves, whom they trust as friends. The stakeholder's logic behind listening to trusted friends relates to the risk and uncertainty of adopting new technology. *Structural equivalence* explains diffusion as a copycat approach. The decision to adopt is not based on sound judgment, but through fear and risk adversity. The last concept, *thresholds*, states that diffusion is a complex process that can be influenced by education, wealth, communication networks, and background. An innovation is not diffused over homogenous people, but between diverse individuals with different backgrounds. According to the concept of thresholds, a stakeholder's decision to adopt a new technology is interconnected with other stakeholders.[55]

5.2. Factors of Diffusion

Some innovations, such as cellular phones, only take a few years to reach widespread adoption, while others can take decades. Characteristics of innovations can provide some explanation for this difference. Rogers identifies five primary characteristics as seen in Figure 5.3: relative advantage, compatibility, complexity, trialability, and observability. The relative advantage is the extent that an innovation is perceived to be better than the current or previous idea. Compatibility is the extent that a new innovation is consistent with current values and needs. Innovations that are compatible with current norms and needs are likely to be adopted more rapidly than one that is not compatible. Complexity refers to the perception of how complicated a new innovation is to understand and use. Increased complexity slows the adoption of a new innovation. Trialability is the extent that a new innovation may be tested before fully adopting it. Observability is the extent that the use and results of a new innovation can be seen by would-be adopters.

I. Perceived Attributes of Innovations
   i. Relative advantage
   ii. Compatibility
   iii. Complexity
   iv. Trialability
   v. Observability
II. Type of Innovation-Decision
   i. Optional
   ii. Collective
   iii. Authority
III. Communication Channels
IV. Nature of the Social System
V. Extent of Change Agents' Promotion Efforts

Rate of Adoption of Innovations

Rogers, E. M. (1995). *Diffusion of Innovations,* 5th Edition (New York: The Free Press, 2003), 222.

Figure 5.3. Variables Determining the Rate of Adoption of Innovations.

The type of innovation decision is also a factor in the rate of adoption. Optional innovation decisions are decisions made by individuals independent of other members of a system; thus, the individual is the main unit of decision making. Collective innovation decisions are those decisions that are made by consensus among members of a system. Authority innovation decisions are those decisions to adopt or reject an innovation by a select few individuals who maintain power, status, or technical expertise. For example, a chief executive officer (CEO) who decides that all employees will wear a suit would be an authority decision.

A communication channel, as referred to in Figure 5.3, is the means by which individuals communicate concerning an innovation. These might include the evaluation of an innovation by a peer or a review by an expert.

One-on-one and other communications often take place within a social system. Communication may also occur through mass media. Communication channels are important in determining the diffusion of an innovation; however, it often requires in-depth investigation to understand these channels.

The nature of the social system, such as its norms and interconnectedness, is also an important factor in the diffusion of an innovation. This includes the system's culture, but also includes the network of connections between potential adopters. This can be a significant factor in the diffusion of a technology, especially in the case where the preferred communication channel is one-on-one interaction. Similar to the communication channels, the nature of the social system is an important factor, but this type of information is not well documented. Additional research may be needed to develop a full understanding of both the social system and relevant communication channels.

The last variable is the change agent. Both public and private organizations strive to change the marketplace. Many entities provide incentives or subsidies in order to speed up the rate of adoption of innovations. For example, the federal government often creates incentives for individuals or businesses to adopt more environmentally friendly products such as energy efficient lighting. Other events, organizations, people, or items also act as a catalyst for change in an industry.

5.3. Diffusion of Additive Manufacturing

Globally, 6494 industrial additive manufacturing systems were deployed in 2011 with a cumulative total of 49 035 systems being deployed between 1988 and 2011. Of these, 18 780 were deployed in the U.S. The growth in the cumulative number of additive manufacturing systems in the U.S. between 2010 and 2011 was 15.3%.[56]

The status of some of the variables that affect the adoption of additive manufacturing technologies can be observed through existing articles and texts; however, many issues cannot be substantiated without gathering additional data. Surveys can often be used to assess a producer or user's opinion of a new technology, but this is often a resource intensive process. Using the number of domestic unit sales[57], the growth in sales can be fitted using least squares criterion to an exponential curve that represents the traditional logistic S-curve of technology diffusion. The most widely accepted model of technology diffusion was presented by Mansfield[58]:

$$p(t) = \frac{1}{1 + e^{\alpha - \beta t}}$$

Where

$p(t)$ = the proportion of potential users who have adopted the new technology by time t

α = Location parameter

β = Shape parameter ($\beta > 0$)

In order to examine additive manufacturing, it is assumed that the proportion of potential units sold by time t follows a similar path as the proportion of potential users who have adopted the new technology by time t. In order to examine shipments in the industry, it is assumed that an additive manufacturing unit represents a fixed proportion of the total revenue; thus, revenue will grow similarly to unit sales. The proportion used was calculated from 2011 data. The variables α and β are estimated using regression on the cumulative annual sales of additive manufacturing systems in the U.S. between 1988 and 2011. U.S. system sales are estimated as a proportion of global sales. This method provides some insight into the current trend in the adoption of additive manufacturing technology. Unfortunately, there is little insight into the total market saturation level for additive manufacturing; that is, there is not a good sense of what percent of the relevant manufacturing industries (shown in Table 4.1) will produce parts using additive manufacturing technologies versus conventional technologies. In order to address this issue, a modified version of Mansfield's model is adopted from Chapman[59]:

$$p(t) = \frac{\eta}{1 + e^{\alpha - \beta t}}$$

where

η = market saturation level

Because η is unknown, it is varied between 0.15% and 100% of the relevant manufacturing shipments, as seen in Table 5.1. The 0.15% is derived from Wohlers estimate that the 2011 sales revenue represents 8% market penetration, which equates to \$3.1 billion in market opportunity and 0.15%

market saturation. At this level, additive manufacturing is forecasted to reach 50% market potential in 2018 and 100% in 2045, as seen in the table. A more likely scenario seems to be that additive manufacturing would have between 5% and 35% market saturation. At these levels, additive manufacturing would reach 50% of market potential between 2031 and 2038 while reaching 100% between 2058 and 2065, as seen in Table 5.1. The industry would reach $50 billion between 2029 and 2031 while reaching $100 billion between 2031 and 2044. As illustrated in Figure 5.4, it is likely that additive manufacturing is at the far left tail of the diffusion curve, making it difficult to forecast the future trends; thus, some caution should be used when interpreting this forecast. The figure illustrates the diffusion at each market saturation level presented in Table 5.1 with the exception of the 0.50% and 0.15% levels, as they are too small to be included in this graph.

Table 5.1. Forecasts of U.S. Additive Manufacturing Shipments by Varying Market Potential

Market Potential of Relevant Manufacturing (percent of shipments)	Market Potential, Shipments ($billions 2011)	Approximate Year 100% of Market Potential Reached	Approximate Year 50% of Market Potential Reached	Approximate Year $100 Billion in Shipments is Reached	Approximate Year $50 Billion in Shipments is Reached	R2
100.00	$2 058.9	2069	2042	2031	2028	0.948
75.00	$1 544.2	2068	2041	2031	2028	0.948
50.00	$1 029.5	2067	2039	2031	2029	0.948
45.00	$926.5	2066	2039	2031	2029	0.948
40.00	$823.6	2066	2038	2031	2029	0.948
35.00	$720.6	2065	2038	2031	2029	0.948
30.00	$617.7	2065	2037	2031	2029	0.948
25.00	$514.7	2064	2037	2032	2029	0.948
20.00	$411.8	2063	2036	2032	2029	0.948
15.00	$308.8	2062	2035	2032	2029	0.948
10.00	$205.9	2061	2033	2033	2029	0.948
5.00	$102.9	2058	2031	2044	2031	0.948
1.00	$20.6	2052	2025	-	-	0.949
0.50	$10.3	2050	2023	-	-	0.949
0.15	$3.1	2045	2018	-	-	0.950

5.3.1. Perceived Attributes of Innovation

Relative Advantage: The relative advantage of adopting additive manufacturing varies from industry to industry and is likely to increase over time as the technology advances. The per-unit cost of additive manufacturing appears to be a significant barrier for many would-be adopters. For some, the benefits outweigh the costs. For instance, lighter transportation equipment can significantly reduce costs for end users; thus, they might be willing to pay higher upfront costs to purchase lighter equipment made using additive manufacturing technologies. For others, however, the benefits of products made using this technology may not justify the higher costs for producers or end users. One possible challenge that could develop is communicating and convincing the end user of the benefits of a product made using additive manufacturing. For instance, this technology may allow for the design of a longer lasting product; however, the end user is only willing to pay for the additional costs of production if they are aware of and convinced of the benefits.

One of the primary beneficiaries of additive manufacturing is the end user; thus, their role in persuading manufacturers to adopt additive manufacturing technology is a significant one. On the other hand, manufacturers may need to differentiate products made using additive manufacturing technology by indicating the benefits to the end user; otherwise, costumers may not be willing to pay the costs for these products.

Compatibility: The limited size of the products that can be produced affects the compatibility of additive manufacturing for some manufactured products. Transportation equipment, for instance, involves large components that may be difficult to produce using additive manufacturing technology.

Complexity, Trialability and Observability: Additive manufacturing systems can be costly; however, these systems are seemingly easy to illustrate and a significant amount of literature is available on them. Currently, there are journals and conferences that discuss this technology extensively. One challenge that seems to persist is cost categorization and analysis. This prevents a prospective manufacturer from observing the costs and benefits from adopting this technology. A number of developments have been made on this front; however, no model meets all criteria adequately. There is a need to bring together the strengths of existing cost models into one standardized model.[60] This would allow would-be adopters to understand the benefits and costs more adequately.

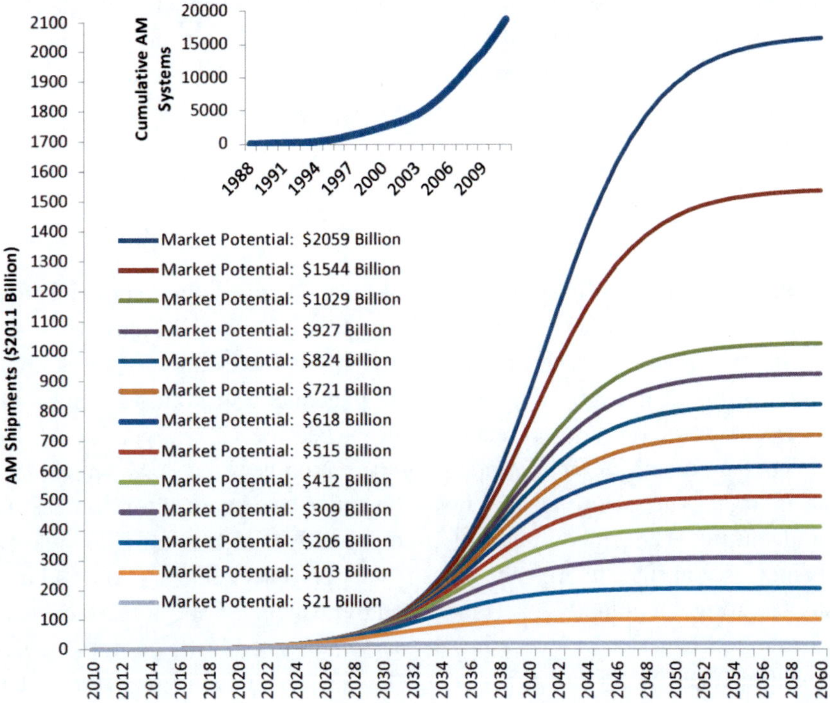

Figure 5.4. Forecasts of U.S. Additive Manufacturing Shipments, by Varying Market Saturation Levels.

5.3.2. Change Agents

The last factor involves the efforts of change agents. These entities can be individuals, events, organizations, or some other entity that acts as a catalyst for change. They often accelerate the realization of benefits, reduce costs, and/or increase benefits of some trend in society or the economy. This change can often occur through research and collaboration efforts. For additive manufacturing, there are a number of organizations that strive to advance the current status. One newly created organization is the National Additive Manufacturing Innovation Institute (NAMII), which was formally established in 2012 with an initial $30 million in federal funding matched by $40 million from a consortium of companies, universities, colleges, and non-profit organizations. The single focus of NAMII is to "accelerate additive manufacturing technologies to the U.S. manufacturing sector and increase domestic manufacturing competitiveness."[61] Likewise, the Additive Manufacturing Consortium (AMC) was launched by EWI. The mission of the

AMC is to "bring together a diverse group of practitioners and stakeholders that together accelerate the innovation in AM technologies to move them into the mainstream of manufacturing technology from their present emerging position."[62]

6. Opportunities for Change Agents

Metrics used to discuss national industries often involve examining the amount of research being conducted, factors that impact the industry, or the size of the industry. The primary purpose of investing resources into manufacturing activities, however, is to generate a benefit or return on investment. Arguably, those countries that exceed the U.S. in per capita size and benefits per unit of input, such as compensation per hour, have an industry that is more successful at the main objective of investing resources in manufacturing. The general purpose of an industry change agent is to create a net increase in the return on investment for stakeholders. For additive manufacturing, this might be accomplished by reducing costs, accelerating the realization of benefits, or increasing the net benefits as illustrated in the larger graph illustrated in Figure 6.1. These changes result in an increase in the marginal return on investment as illustrated in the smaller graph in the figure.

Generally, change agents want to maximize their impact for the amount of resources allotted to them; that is, they want the "biggest bang for the buck." Investment in any particular change agent effort, traditionally, has decreasing returns to scale; that is, every additional dollar of investment has a little less impact than the previous dollar. Since a change agent wants to maximize their impact, it would want to allocate its funding in projects such that each dollar of investment has the maximum return possible. For instance, Figure 6.2 provides an illustration of five possible investments for a change agent with a budget constraint. The investments are referred to as efforts A through E. To maximize its impact, a change agent would first invest in Effort A. As it invests more and more in Effort A, it moves from the left to the right along the marginal return on investment line for the change agent. The agent would invest to the point where the marginal return on investment for its next dollar invested equals that of Effort B, which is referred to in the figure as the "Point at which B becomes worthwhile." At this point, there is some indifference to investing in A or B because they have the same marginal return on investment; however, as one invests in either A or B the return on investment in that effort decreases making the alternative more appealing. Therefore, the change agent would invest in both A and B or alternate between the two until the point at which the next effort becomes worthwhile. It would continue to do this until its entire budget is expended. In this example, effort E goes unfunded while efforts A through D are funded to where the bottom of each corresponding green line stops; thus, the total investment is the sum of the investment level for Effort A, B, C, and D. It is important to note, however, that not all of the costs and benefits of the manufacturing industry are able to be measured nor are the impact of the efforts of change agents; therefore, identifying the optimal use of funding can be rather problematic.

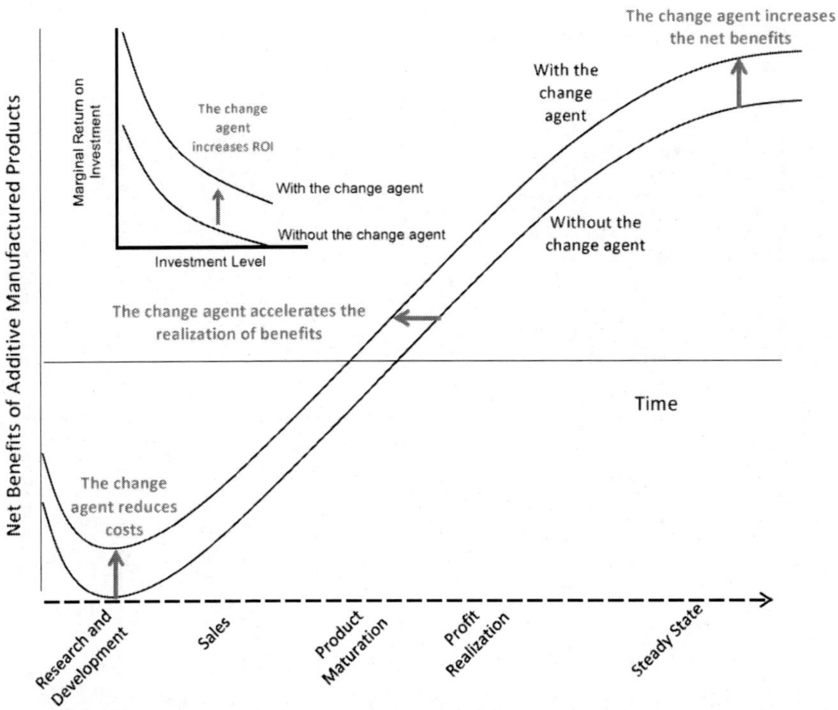

Modified from Gallaher, Michael P., Thomas Phelps and Alan C. O'Connor. Planning Report 02-5: Economic Impact Assessment of the International Standard for the Exchange of Product Model Data (STEP) in Transportation Equipment Industries. RTI International and the National Institute of Standards and Technology. December 2002: 5-4.

Figure 6.1. Impact of Change Agents on the Net Benefits and Return on Investment for Additive Manufactured Products.

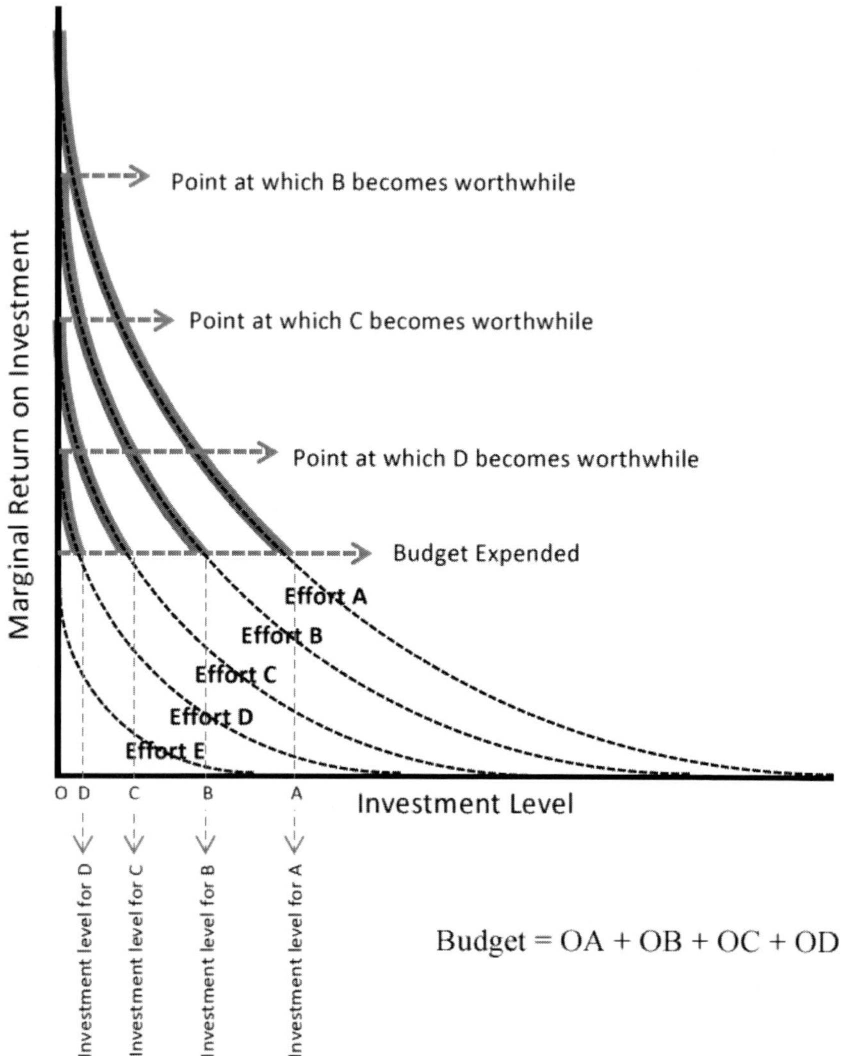

Note: The green lines represent the investment for each effort.

Figure 6.2. Illustration of the Optimal use of Change Agent Funding for Six Alternative Investments.

Change agents for the additive manufacturing industry can focus their efforts on three primary areas to advance this technology: cost reduction, accelerating the realization of benefits, and increasing the benefits of additive manufacturing. The costs include any of the investments made by the stakeholders listed in Table 3.2. These include the owners, employees, suppliers, and end users among others. The producer costs of additive manufacturing tend to be broken into preparation, materials, machine utilization, and post processing. As seen in some case studies in Section 4.3, the largest cost tends to be the machine operation cost followed by the material cost. The time it takes to produce a product may be a significant factor in the machine utilization cost. Since these two costs are the largest, there is a potentially high marginal return on investment for change agents that focus on reducing these costs; that is, focusing on these items may result in a higher return on investment for some change agents. However, examining these issues in detail provides some challenge as there is not a standard cost categorization. This prevents change agents from precisely identifying the major costs of this technology. A number of developments have been made on this front; however, no model meets all criteria adequately. There is a need to bring together the strengths of existing cost models into one standardized model.[63] This might be another area that has a high return on investment for change agents.

As previously discussed, a major benefit of this technology is in the area of product design and how it allows the production of nearly any complexity of geometry without the need for tooling. Additionally, the complexity does not impact the cost in the same way that it does for conventional manufacturing. This technology eliminates many of the restrictions of 'Design for Manufacture and Assembly' opening a new realm of possibilities for new customized products at an affordable price point. To some degree, the success of this technology will rely on taking advantage of this benefit. In order to achieve this, the products must meet quality and reliability standards and there must be testing standards in place to verify their performance. For instance, the U.S. Federal Aviation Regulations have strict regulations for material performance related to fatigue, creep, flammability, and toxicity. Manufacturers rely on standards in materials and processes to ensure the performance of their products.[64] The dissimilarities between conventional manufacturing processes and those of additive manufacturing are likely to require modifications to current performance validation processes.[65] Standards and codes organizations will likely play a significant role in facilitating the adoption of additive manufacturing technology.

Although this technology can produce nearly any complexity of geometry, it is limited in the size of the components that can be constructed. Expanding the size while maintaining a reasonable price point is likely to increase the rate at which this technology is adopted and expand the market opportunity.

Additionally, the quality of the product is a limiting factor. Materials or surface finish, for instance, can often be inadequate for parts and components.

CONCLUSION

There is a general concern that the U.S. manufacturing industry has lost competitiveness with other nations; however, industry data suggests that the U.S. still maintains a prominent position. Additive manufacturing may provide an important opportunity for advancing U.S. manufacturing while maintaining and advancing U.S. innovation. The U.S. is currently a major user of additive manufacturing technology and the primary producer of additive manufacturing systems. Globally, an estimated $642.6 million in revenue was collected for additive manufactured goods, with the U.S. accounting for an estimated $246.1 million or 38.3% of global production in 2011. Approximately 62.8% of all commercial/industrial units sold in 2011 were made by the top three producers of additive manufacturing systems: Stratasys, Z Corporation, and 3D Systems based out of the U.S. Approximately 64.4% of all systems were made by companies based in the U.S. If additive manufacturing has a saturation level between 5% and 35% of the relevant sectors, it is forecasted that it might reach 50% of market potential between 2031 and 2038, while reaching 100% between 2058 and 2065, as seen in Table 5.1. The industry would reach $50 billion between 2029 and 2031, while reaching $100 billion between 2031 and 2044. Since it is likely that additive manufacturing is at the far left tail of the diffusion curve, making it difficult to forecast the future trends, some caution should be used when interpreting these estimates.

Change agents for the additive manufacturing industry can focus their efforts on three primary areas: costs, rate at which benefits are realized, or the benefits of additive manufacturing. Costs have been identified as being a significant factor in whether producers adopt additive manufacturing technologies. Hopkinson[66] estimates that machine costs range between 50% and 75% of total cost, materials range between 20% and 40%, and labor ranges between 5% and 30%. Reducing these costs may have a significant impact on the adoption of additive manufacturing technologies. Additionally, quality, performance validation, and expanding size capabilities are likely to also have significant impacts.

REFERENCES

Atzeni, Eleonora, Luca Iuliano, Paolo Minetola, and Alessandro Salmi. (2010) "Redesign and Cost Estimation of Rapid Manufactured Plastic Parts." Rapid Prototyping Journal. 16(5): 308-317.

Boothroyd, Geoffrey, Peter Dewhurst, and Winston Knight. Product Design for Manufacture and Assembly. (New York: Marcel Dekker, Inc, 2009).

Bourell, David L., Ming C. Leu, and David W. Rosen. "Roadmap for Additive Manufacturing: Identifying the Future of Freeform Processing." University of Texas. <http://wohlersassociates.com/roadmap2009.html>

Chapman, Robert. "Benefits and Costs of Research: A Case Study of Construction Systems Integration and Automation Technologies in Commercial Buildings." NISTIR 6763. December 2001. National Institute of Standards and Technology.

Davidson, Adam. "Making It in America." The Atlantic. January/February (2012). <http://www.theatlantic.com/magazine/archive/2012/01/making-it -in-america/8844/?singlejage=true>

Davidson, Adam. "The Transformation of American Factory Jobs, In One Company." NPR. January 13, 2012. <http://www.npr.org/blogs/money /2012/01/13/145039131/the-transformation-of-american-factoryjobs-in-one-company?ft=1&f=100>

Economist. "Printing Body Parts: Making a Bit of Me." <http://www. economist.com/node/15543683>

Gausemeier, Jurgen, Niklas Echterhoff, Martin Kokoschika, and Marina Wall. "Thinking Ahead the Future of Additive Manufacturing – Future Applications." University of Paderborn, Direct Manufacturing Research Center.

Gibson, Ian, David Rosen, and Brent Stucker. Additive Manufacturing Technologies. Springer: New York, 2010. 47-50

GizMag. "3D Bio-printer to Create Arteries and Organs." <http://www. gizmag.com/3d-bio-printer/13609/>

Greenwald, Bruce C.N. and Judd Kahn. Globalization: The Irrational Fear that Someone in China will Take Your Job. (Hoboken, NJ: John Wiley & Sons 2009).

Hopkinson, Neil, "Production Economics of Rapid Manufacture." In Hopkinson, Neil, Richard Hague, and Philip Dickens. Rapid Manufacturing. (Hoboken, NJ: John Wiley & Sons, 2006). 147-157.

Horowitz, Karen J. and Mark A. Planting. Concepts and Methods of the U.S. Input-Output Accounts. Bureau of Economic Analysis. 2006.

Koebel, C. Theodore, Maria Papadakis, Ed Hudson, Marilyn Cavell, The Diffusion of Innovation in the Residential Building Industry, PATH, p. 1.

Krugman, Paul R. “Competitiveness, A Dangerous Obsession.” Foreign Affairs. Vol 73. Num 2. March/April (1994): 28-44.

Krugman, Paul R. “Making Sense of the Competitiveness Debate.” Oxford Review of Economic Policy. Vol 12, no. 3 (1996): 17-25. Paul Krugman won the 2008 Nobel Memorial Prize in Economic Sciences for his work on international trade and economic geography.

Larsen, Graeme D., “Horses for Courses: Relating Innovation Diffusion Concepts to the Stages of the Diffusion Process,” Construction Management and Economics, Vol 23, October 2005, p. 787-792.

Lindemann C., U. Jahnke, M. Moi, and R. Koch. “Analyzing Product Lifecycle Costs for a Better Understanding of Cost Drivers in Additive Manufacturing.” Proceedings of the 2012 Solid Freeform Fabrication Symposium. <http://utwired.engr.utexas.edu/lff/symposium/proceedings Archive/pubs/Manuscripts/2012/2012-12- Lindemann.pdf>

Mansfield, Edwin. Innovation, Technology and the Economy: Selected Essays of Edwin Mansfield. Economists of the Twentieth Century Series (Brookfield, VT: 1995, E. Elgar).

Mansour, S., Richard Hague. (2003) “Impact of Rapid Manufacturing on Design for Manufacture for Injection Molding.” Proceedings of the Institution of Mechanical Engineers, Part B: Journal of Engineering Manufacture.

McKinsey&Company. “Manufacturing the Future: The Next Era of Global Growth and Innovation.” November 2012. <http://www.mckinsey.com /insights/mgi/research/productivity_competitiveness_and_growth/the_future_of _manufacturing>

National Academy of Engineering. “Frontiers of Engineering 2011: Reports on Leading-Edge Engineering from the 2011 Symposium.” In National Academy of Engineering’s 2011 U.S. Frontiers of Engineering Symposium. Mountain View, CA. 2012

National Institute of Standards and Technology. “Roadmapping Workshop: Measurement Science for Metal-Based Additive Manufacturing.” <http://events.energetics.com/nistadditivemfgworkshop/index.html>

National Science Foundation. “Asia’s Rising Science and Technology Strength.” May 2007. <http://www.nsf.gov/statistics/nsf07319/>

OECD (2012), OECD Science, Technology and Industry Outlook 2012, OECD Publishing. <http://dx.doi.org/10.1787/sti_outlook-2012-en>

Porter, Michael E. “Building the Microeconomic Foundations of Prosperity: Findings from the Business Competitiveness Index.” In Porter, Michael E., Klaus Schwab, Xavier Sala-i-Martin, and Augusta LopezClaros. The Global Competitiveness Report 2003-2004. (New York: Oxford University Press, 2004).

Porter, Michael E. The Competitive Advantage of Nations. 1st ed. (New York: The Free Press, 1990).

Rogers, E. M. (2003). Diffusion of Innovations, Fourth Edition (New York: The Free Press, 2003), p. 111- 114.

Scott, Justin, Nayanee Gupta, Christopher Weber, Sherrica Newsome, Terry Wohlers, and Tim Caffrey. “Additive Manufacturing: Status and Opportunities”, March 2012. <https://www.ida.org/stpi/occasionalpapers/papers/AM3D_33012_Final.pdf>

Sirkin, Harold L. “Made in the USA Still Means Something.” Bloomberg Businessweek. April 10, 2009. <http://www.businessweek.com/managing/content/apr2009/ca20090410_054122.htm>

Slaughter, Matthew J. “How U.S. Multinational Companies Strengthen the U.S. Economy.” United States Council for International Business. (March 2010). <http://www.uscib.org/docs/foundation_multinationals.pdf>

Tassey Gregory. “Rationales and Mechanisms for Revitalizing U.S. Manufacturing R&D Strategies.” Journal of Technology Transfer. 35 (2010): 283-333.

Thomas, Douglas S. “The Current State and Recent Trends of the U.S. Manufacturing Industry”, NIST Special Publication 1142. December 2012. <http://www.nist.gov/manuscript-publicationsearch.cfm?pub_id=912933>

Thomas, Douglas. “National Industry Performance Metrics: A Case Study of U.S. Manufacturing.” National Institute of Standards and Technology. White paper. Available upon request.

Thomson Reuters. “Top 100 Global Innovators, 2011.” <http://www.top100innovators.com/overview>

Triadic patent families are defined at the OECD as a set of patents taken at the European Patent Office, Japanese Patent Office, and U.S. Patent and Trademark Office that share one or more priorities.

Vishwanath, Arun and George Barnett. The Diffusion of Innovations. (New York: Peter Lang, 2011).

West, Karl. “Melted Metal Cuts Plane’s Fuel Bill.” The Sunday Times. Sunday 13 February 2011. <http://www.thesundaytimes.co.uk/sto/business/energy_and_environment/article547163.ece>

Wohlers, Terry. "Wohlers Report 2012: Additive Manufacturing and 3D Printing State of the Industry." Wohlers Associates, Inc. 2012: 130.

World Economic Forum. The Global Competitiveness Report. 2010-2011. <http://www3.weforum.org/docs/WEF_GlobalCompetitivenessReport_2010-11.pdf>

APPENDIX A: SCHEMATIC DATA MAP

The Annual Survey of Manufactures (ASM) is conducted every year except for years ending in 2 or 7 when the Economic Census is conducted. The ASM provides statistics on employment, payroll, supplemental labor costs, cost of materials consumed, operating expenses, value of shipments, value added, fuels and energy used, and inventories. It uses a sample survey of approximately 50 000 establishments with new samples selected at 5- year intervals. An establishment is an economic unit—business or industrial—at a single physical location where business is conducted or where services or industrial operations are performed. The ASM data allows the examination of multiple factors (value added, payroll, energy use, and more) of manufacturing at a detailed subsector level. The Economic Census, used for years ending in 2 or 7, is a survey of all employer establishments in the U.S. that has been taken as an integrated program at 5-year intervals since 1967. Both the ASM and the Economic Census use NAICS classification; however, prior to NAICS the Standard Industrial Classification system was used. Table A.1 contains items from the Annual Survey of Manufactures. The color scheme matches that of the color scheme in the manufacturing supply chains presented previously in this report.

Each supply chain item is calculated for the NAICS codes listed in Table 4.1 and added together by the categories listed in the table using data from the Annual Survey of Manufactures seen in Table A.2. The values for additive manufacturing seen in Table A.3 are calculated by assuming that the ratio of each supply chain item to the total value of shipments from the data in Table A.2 is the same for additive manufacturing. The ratios are then applied to data in the 2012 Wohlers Report. These assumptions have significant implications for precision; however, they are the best estimates available.

Table A.1. Supply Chain Components

ASM Data Item	Schematic name
Number of employees	Payroll, Benefits, and employment
Annual payroll	Payroll, Benefits, and employment
Total fringe benefits	Payroll, Benefits, and employment
Employer's cost for health insurance	
Employer's cost for defined benefit pension plans	
Employer's cost for defined contribution plans	
Employer's cost for other fringe benefits	
Production workers avg per year	
Production workers hours (1,000)	
Production workers wages	
Total cost of materials	
Materials, parts, containers, packaging, etc. used	Materials, parts, containers, packaging, etc used
Cost of resales	Contract work and resales
Contract work	Contract work and resales
Cost of purchased fuels	Purchased fuels and electricity
Purchased electricity	Purchased fuels and electricity
Quantity of electricity purchased	
Quantity of generated electricity	
Quantity of electricity sold or transferred	
Total value of shipments	Shipments
Value of products shipments	
Total miscellaneous receipts	
Value of resales	
Contract receipts	
Other miscellaneous receipts	
Value of interplant transfers	
Value added	Value added
Total EOY inventories	
Finished goods inventories, EOY	Net Inventories Shipped
Work-in-process inventories, EOY	Net Inventories Shipped
Materials and supplies inventories, EOY	

Table A.1. (Continued)

Total BOY inventories	
Finished goods inventories, BOY	Net Inventories Shipped
Work-in-process inventories, BOY	Net Inventories Shipped
Materials and supplies inventories, BOY	
Total capital expenditures (new and used)	
Capital expenditures: buildings & other structures (new and used)	Capital expenditures: buildings and other structures (new and used)
Capital expenditures: machinery and equipment (new and used)	
Capital expenditures: autos, trucks, etc. for highway use	Capital expenditures: machinery and equipment (new and used)
Capital expenditures: computer and data processing equipment	Computer hardware, software, and other equipment
Capital expenditures: all other machinery and equipment	Capital expenditures: machinery and equipment (new and used)
Total depreciation	Depreciation
Total rental payments	
Buildings rentals	Capital expenditures: buildings and other structures (new and used): Rental
Machinery rentals	Capital expenditures: machinery and equipment (new and used): Rental
Total other expenses	
Temporary staff and leased employee expenses	Other costs
Expensed computer hardware and other equipment	Computer hardware, software, and other equipment
Expensed purchases of software	Computer hardware, software, and other equipment
Data processing and other purchased computer services	Professional, technical, and data services
Communication services	Communication services
Repair and maintenance services of buildings and/or machinery	Maintenance and repair
Refuse removal (including hazardous waste) services	Refuse removal
Advertising and promotional services	Other costs
Purchased professional and technical services	Professional, technical, and data services
Taxes and license fees	Other costs
All other expenses	Other costs

Volume of Production=total costs (blue plus orange plus red plus green plus gold)

Net Inventories Shipped=sum of EOY finished goods and work-in-process inventories less the sum of BOY finished goods and work-in-process inventories

Table A.2. Total Supply Chain Values for Industries Relevant to Additive Manufacturing, $million 2011

	Communication Services	Other Costs	Refuse Removal	Computer Hardware, Software, and other Equipment	Professional, Technical, and Data Services	Payroll, Benefits, and Employment	Employment	Capital Expenditures:Buildings and Other Structures	Capital Expenditures: Machinery and Equipment	Materials, Parts, Containers, Packageing, etc. Used
Motor vehicles	167	14 995	391	542	1 408	47 238	651	2 345	10 686	307 489
Aerospace	155	7 682	177	689	3 142	34 987	333	1 323	2 515	65 064
Industrial/business machines	528	22 773	475	1 351	2 863	70 427	965	3 802	8 876	169 346
Medical/dental	151	8 903	123	477	1 409	21 533	290	1 230	1 850	22 633
Government/military	52	1 813	57	192	341	11 024	82	380	407	9 482
Architectural	121	4 823	101	226	494	17 631	302	1 090	1 469	32 747
Consumer products/electronics, academic institutions, and other	1 854	64 203	1 629	4 540	7 819	205 764	2 894	11 589	26 634	338 544
Total	3 028	125 192	2 952	8 016	17 477	408 603	5 516	21 760	52 437	945 304

Table A.2. (Continued)

	Contract Work and Resales	Purchased Fuels and Electricity	Maintenance and Repair	Volume of Production	Net Inventories Shipped	Depreciation	Net Income	Shipments	Value Added (ASM)	*Additive Manufacturing's Share of Shipments*
Motor vehicles	9 300	2 806	2 116	399 482	-1 108	9 695	37 220	445 289	126 751	*0.021%*
Aerospace	7 739	1 257	611	124 628	-5 920	2 181	36 812	157 701	90 216	*0.036%*
Industrial/business machines	20 494	2 704	2 218	305 856	-4 414	6 477	57 815	365 735	177 486	*0.014%*
Medical/dental	4 446	525	450	63 730	-17	1 987	23 819	89 519	61 932	*0.079%*
Government/military	3 386	221	117	29 132	-323	468	3 507	32 784	18 350	*0.086%*
Architectural	4 934	723	441	64 799	-408	1 458	6 338	72 187	34 162	*0.019%*
Consumer products/electronics, academic institutions, and other	48 869	9 592	6 319	726 410	-5 886	22 966	152 219	895 710	505 513	*0.018%*
Total	99 168	17 828	12 273	1 714 038	-18 076	45 234	317 730	2 058 926	1 014 411	*0.023%*

Table A.3. Supply Chain Values for Additive Manufacturing by Industry, $million 2011

	Communication Services	Other Costs	Refuse Removal	Computer Hardware, Software, and other Equipment	Professional, Technical, and Data Services	Payroll, Benefits, and Employment	Employment	Capital Expenditures: Buildings and Other Structures	Capital Expenditures: Machinery and Equipment	Materials, Parts, Containers, Packageing, etc. Used
Motor vehicles	0.02	1.6	0.04	0.1	0.2	5.1	70	0.3	1.2	33.1
Aerospace	0.03	1.5	0.03	0.1	0.6	6.6	63	0.2	0.5	12.3
Industrial/business machines	0.04	1.7	0.03	0.1	0.2	5.1	70	0.3	0.6	12.3
Medical/dental	0.06	3.7	0.05	0.2	0.6	8.9	120	0.5	0.8	9.4
Government/military	0.02	0.8	0.03	0.1	0.2	5.0	37	0.2	0.2	4.3
Architectural	0.01	0.5	0.01	0.0	0.1	1.8	31	0.1	0.2	3.3
Consumer products/electronics, academic institutions, and other	0.17	5.9	0.15	0.4	0.7	19.0	267	1.1	2.5	31.3
Total	0.4	15.7	0.3	1.0	2.5	51.5	658.3	2.6	5.8	106.0

Table A.3. (Continued)

	Contract Work and Resales	Purchased Fuels and Electricity	Maintenance and Repair	Volume of Production	Net Inventories Shipped	Depreciation	Net Income	Shipments	Value Added (ASM)	*Additive Manufacturing's Share of total Shipments*
Motor vehicles	1.0	0.3	0.2	43.1	-0.1	1.0	4.0	48.0	13.7	*0.01%*
Aerospace	1.5	0.2	0.1	23.5	-1.1	0.4	7.0	29.8	17.0	*0.02%*
Industrial/business machines	1.5	0.2	0.2	22.2	-0.3	0.5	4.2	26.6	12.9	*0.01%*
Medical/dental	1.8	0.2	0.2	26.5	0.0	0.8	9.9	37.2	25.7	*0.04%*
Government/military	1.5	0.1	0.1	13.1	-0.1	0.2	1.6	14.8	8.3	*0.05%*
Architectural	0.5	0.1	0.0	6.6	0.0	0.1	0.6	7.4	3.5	*0.01%*
Consumer products/electronics, academic institutions, and other	4.5	0.9	0.6	67.1	-0.5	2.1	14.1	82.7	46.7	*0.01%*
Total	12.3	2.0	1.4	202.1	-2.3	5.2	41.3	246.1	127.7	*0.01%*

APPENDIX B: EQUATIONS AND ASSUMPTIONS

The approximations for U.S. additive manufacturing activity rely on the assumption that the U.S. share of additive manufacturing systems sold equates to the share of products produced using additive manufacturing systems. This is represented as the following:

$$R_{US} = \frac{S_{US}}{S_G} R_G$$

Where:

R_{US} = Revenue for additive manufacturing activities in the U.S.

S_{US} = Cumulative number of additive manufacturing systems sold in the U.S. between 1988 and 2001

S_G = Cumulative number of additive manufacturing systems sold globally between 1988 and 2001

R_G = Revenue from the global sale of parts produced from additive manufacturing systems

Shipments of additive manufactured parts and products by category (see Table 4.1) was estimated by assuming that the percent of additive manufacturing that each category represents is the same for the U.S. as it is globally. The calculation is represented as the following:

$$R_{US,x} = \frac{R_{G,x}}{R_G} R_{US}$$

Where:

$R_{US,x}$ = U.S. revenue for additive manufacturing activities for category x

$R_{G,x}$ = Global revenue for additive manufacturing activities for category x

R_G = Global revenue for additive manufacturing

R_{US} = Revenue for additive manufacturing activities in the US

End Notes

[1] Economist. "Printing Body Parts: Making a Bit of Me." <http:// www.economist.com /node/15543683>

[2] GizMag. "3D Bio-printer to Create Arteries and Organs." <http://www. gizmag.com/3d-bioprinter/13609/>

[3] Wohlers, Terry. "Wohlers Report 2012: Additive Manufacturing and 3D Printing State of the Industry." Wohlers Associates, Inc. 2012.

[4] Wohlers, Terry. "Wohlers Report 2012: Additive Manufacturing and 3D Printing State of the Industry." Wohlers Associates, Inc. 2012.

[5] Thomas, Douglas S. "The Current State and Recent Trends of the U.S. Manufacturing Industry", NIST Special Publication 1142. December 2012. <http://www.nist.gov /manuscript-publicationsearch.cfm?pub_id= 912933>

[6] Gibson, Ian, David Rosen, and Brent Stucker. Additive Manufacturing Technologies. Springer: New York, 2010. 47-50

[7] Davidson, Adam. "The Transformation of American Factory Jobs, In One Company." NPR. January 13, 2012. <http://www.npr.org/blogs/money /2012/01/13/145039131/the-transformation-of-american-factoryjobs-in-one-company?ft=1&f=100>

[8] Davidson, Adam. "Making It in America." *The Atlantic*. January/February (2012). <http://www. theatlantic.com/magazine/archive/2012/01/making-it-in-america/8844/?singlejage=true>

[9] Tassey Gregory. "Rationales and Mechanisms for Revitalizing U.S. Manufacturing R&D Strategies." *Journal of Technology Transfer*. 35 (2010): 283-333.

[10] Slaughter, Matthew J. "How U.S. Multinational Companies Strengthen the U.S. Economy." United States Council for International Business. (March 2010). <http://www.uscib.org /docs/foundation_multinationals.pdf>

[11] National Science Foundation. "Asia's Rising Science and Technology Strength." May 2007. <http://www.nsf.gov/statistics/nsf07319/>

[12] Sirkin, Harold L. "Made in the USA Still Means Something." Bloomberg Businessweek. April 10, 2009. <http://www.businessweek.com/managing/ content/apr2009/ca20090410_ 054122.htm>

[13] Krugman, Paul R. "Making Sense of the Competitiveness Debate." Oxford Review of Economic Policy. Vol 12, no. 3 (1996): 17-25. Paul Krugman won the 2008 Nobel Memorial Prize in Economic Sciences for his work on international trade and economic geography.

[14] Krugman, Paul R. "Competitiveness, A Dangerous Obsession." *Foreign Affairs*. Vol 73. Num 2. March/April (1994): 28-44.

[15] The World Economic Forum defines competitiveness of a nation as "the set of institutions, policies, and factors that determine the level of productivity of a country." This definition relates to productivity and is not consistent with the idea of countries competing for market share. World Economic Forum. *The Global Competitiveness Report*. 2010-2011. <http://www3.weforum.org/docs/WEF_GlobalCompetitivenessReport_2010-11.pdf>

[16] Porter, Michael E. *The Competitive Advantage of Nations*. 1st ed. (New York: The Free Press, 1990).

[17] Porter asserts that competitiveness is measured by productivity and that measuring a country's competitiveness as its share of world markets is "deeply flawed." Porter, Michael E. "Building the Microeconomic Foundations of Prosperity: Findings from the Business Competitiveness Index." In Porter, Michael E., Klaus Schwab, Xavier Sala-i-Martin, and Augusta Lopez-Claros. The Global Competitiveness Report 2003-2004. (New York: Oxford University Press, 2004).

[18] Greenwald, Bruce C.N. and Judd Kahn. Globalization: The Irrational Fear that Someone in China will Take Your Job. (Hoboken, NJ: John Wiley & Sons 2009).

[19] Thomas, Douglas. "National Industry Performance Metrics: A Case Study of U.S. Manufacturing." National Institute of Standards and Technology. White paper. Available upon request.

[20] Triadic patent families are defined at the OECD as a set of patents taken at the European Patent Office, Japanese Patent Office, and U.S. Patent and Trademark Office that share one or more priorities.

[21] OECD (2012), OECD Science, Technology and Industry Outlook 2012, OECD Publishing. <http://dx.doi.org/10.1787/sti_outlook-2012-en>

[22] Thomas, Douglas S. "The Current State and Recent Trends of the U.S. Manufacturing Industry", NIST Special Publication 1142. December 2012. <http://www.nist.gov/ manuscript-publicationsearch.cfm?pub_id=912933>

[23] Thomson Reuters. "Top 100 Global Innovators, 2011." <http://www.top100innovators.com /overview>

[24] Ibid

[25] Hopkinson, Neil, "Production Economics of Rapid Manufacture." In Hopkinson, Neil, Richard Hague, and Philip Dickens. *Rapid Manufacturing*. (Hoboken, NJ: John Wiley & Sons, 2006). 147-157.

[26] Boothroyd, Geoffrey, Peter Dewhurst, and Winston Knight. *Product Design for Manufacture and Assembly*. (New York: Marcel Dekker, Inc, 2009).

[27] Mansour, S., Richard Hague. (2003) "Impact of Rapid Manufacturing on Design for Manufacture for Injection Molding." *Proceedings of the Institution of Mechanical Engineers, Part B: Journal of Engineering Manufacture*.

[28] McKinsey&Company. "Manufacturing the Future: The Next Era of Global Growth and Innovation." November 2012. <http://www.mckinsey.com /insights/mgi/research/productivity_competitiveness_and_growth/the_futu rex_of _manufacturing>

[29] Horowitz, Karen J. and Mark A. Planting. *Concepts and Methods of the U.S. Input-Output Accounts*. Bureau of Economic Analysis. 2006.

[30] Wohlers, Terry. "Wohlers Report 2012: Additive Manufacturing and 3D Printing State of the Industry." Wohlers Associates, Inc. 2012: 129.

[31] This value is calculated with the assumption that the U.S. share of additive manufacturing systems sold equates to the share of products produced using additive manufacturing systems. The share of additive manufacturing systems is available in Wohlers, Terry. "Wohlers Report 2012: Additive Manufacturing and 3D Printing State of the Industry." Wohlers Associates, Inc. 2012: 134.

[32] Wohlers, Terry. "Wohlers Report 2012: Additive Manufacturing and 3D Printing State of the Industry." Wohlers Associates, Inc. 2012: 130.

[33] Each supply chain item is calculated for the NAICS codes listed in Table 4.1 and added together by the categories listed in the table using data from the Annual Survey of Manufactures. The values for additive manufacturing are calculated by assuming that the ratio of each supply chain item to the total value of shipments is the same for additive manufacturing. The ratios are then applied to data in the 2012 Wohlers Report. These assumptions have significant implications for precision; however, they are the best estimates available.

[34] Gausemeier, Jurgen, Niklas Echterhoff, Martin Kokoschika, and Marina Wall. "Thinking Ahead the Future of Additive Manufacturing – Future Applications." University of Paderborn, Direct Manufacturing Research Center.

[35] Wohlers, Terry. "Wohlers Report 2012: Additive Manufacturing and 3D Printing State of the Industry." Wohlers Associates, Inc. 2012: 130.

[36] Bourell, David L., Ming C. Leu, and David W. Rosen. "Roadmap for Additive Manufacturing: Identifying the Future of Freeform Processing." University of Texas. <http:// wohlersassociates.com/roadmap2009.html>

[37] National Institute of Standards and Technology. “Roadmapping Workshop: Measurement Science for Metal-Based Additive Manufacturing.” <http://events.energetics.com/nistadditivemfgworkshop/index.html>

[38] Wohlers, Terry. “Wohlers Report 2012: Additive Manufacturing and 3D Printing State of the Industry.” Wohlers Associates, Inc. 2012: 130.

[39] Bourell, David L., Ming C. Leu, and David W. Rosen. “Roadmap for Additive Manufacturing: Identifying the Future of Freeform Processing.” University of Texas. <http:// wohlersassociates.com/roadmap2009.html>

[40] Gausemeier, Jurgen, Niklas Echterhoff, Martin Kokoschika, and Marina Wall. “Thinking Ahead the Future of Additive Manufacturing – Future Applications.” University of Paderborn, Direct Manufacturing Research Center.

[41] Scott, Justin, Nayanee Gupta, Christopher Weber, Sherrica Newsome, Terry Wohlers, and Tim Caffrey. “Additive Manufacturing: Status and Opportunities”, March 2012. <https:// www.ida.org/stpi/occasionalpapers /papers/AM3D_33012_Final.pdf>

[42] Z Corporation was acquired by 3D Systems Inc. in 2012.

[43] The dollar estimate is assumes that the share of U.S. revenue is equal to the share of U.S. unit sales, which is from Wohlers, Terry. “Wohlers Report 2012: Additive Manufacturing and 3D Printing State of the Industry.” Wohlers Associates, Inc. 2012: 134.

[44] Lindemann C., U. Jahnke, M. Moi, and R. Koch. “Analyzing Product Lifecycle Costs for a Better Understanding of Cost Drivers in Additive Manufacturing.” Proceedings of the 2012 Solid Freeform Fabrication Symposium. <http://utwired.engr.utexas.edu/lff /symposium /proceedingsArchive/pubs/Manuscripts/2012/2012-12- Lindemann.pdf>

[45] West, Karl. “Melted Metal Cuts Plane’s Fuel Bill.” The Sunday Times. Sunday 13 February 2011. <http://www.thesundaytimes.co.uk/sto /business/energy_and_environment/article 547163.ece>

[46] Hopkinson, Neil, “Production Economics of Rapid Manufacture.” In Hopkinson, Neil, Richard Hague, and Philip Dickens. *Rapid Manufacturing*. (Hoboken, NJ: John Wiley & Sons, 2006). 147-157.

[47] Ibid

[48] Lindemann C., U. Jahnke, M. Moi, and R. Koch. “Analyzing Product Lifecycle Costs for a Better Understanding of Cost Drivers in Additive Manufacturing.” Proceedings of the 2012 Solid Freeform Fabrication Symposium. <http://utwired.engr.utexas.edu/lff /symposium_/proceedings Archive/pubs/Manuscripts/2012/2012-12- Lindemann.pdf>

[49] Atzeni, Eleonora, Luca Iuliano, Paolo Minetola, and Alessandro Salmi. (2010) "Redesign and Cost Estimation of Rapid Manufactured Plastic Parts." Rapid Prototyping Journal. 16(5): 308-317.

[50] Koebel, C. Theodore, Maria Papadakis, Ed Hudson, Marilyn Cavell, *The Diffusion of Innovation in the Residential Building Industry*, PATH, p. 1.

[51] Ibid, p. 2.

[52] Vishwanath, Arun and George Barnett. *The Diffusion of Innovations.* (New York: Peter Lang, 2011).

[53] Rogers, E. M. (2003). *Diffusion of Innovations,* Fourth Edition (New York: The Free Press, 2003), p. 111-114.

[54] Larsen, Graeme D., "Horses for Courses: Relating Innovation Diffusion Concepts to the Stages of the Diffusion Process," *Construction Management and Economics*, Vol 23, October 2005, p. 787-792.

[55] Larsen, Graeme D., "Horses for Courses: Relating Innovation Diffusion Concepts to the Stages of the Diffusion Process," *Construction Management and Economics*, Vol 23, October 2005, p. 787-792.

[56] Wohlers, Terry. "Wohlers Report 2012: Additive Manufacturing and 3D Printing State of the Industry." Wohlers Associates, Inc. 2012.

[57] Wohlers, Terry. "Wohlers Report 2012: Additive Manufacturing and 3D Printing State of the Industry." Wohlers Associates, Inc. 2012.

[58] Mansfield, Edwin. *Innovation, Technology and the Economy: Selected Essays of Edwin Mansfield.* Economists of the Twentieth Century Series (Brookfield, VT: 1995, E. Elgar).

[59] Chapman, Robert. "Benefits and Costs of Research: A Case Study of Construction Systems Integration and Automation Technologies in Commercial Buildings." NISTIR 6763. December 2001. National Institute of Standards and Technology.

[60] Lindemann C., U. Jahnke, M. Moi, and R. Koch. "Analyzing Product Lifecycle Costs for a Better Understanding of Cost Drivers in Additive Manufacturing." Proceedings of the 2012 Solid Freeform Fabrication Symposium. <http://utwired.engr.utexas.edu/lff/symposium /proceedings Archive/pubs/Manuscripts/2012/2012-12- Lindemann.pdf>

[61] National Additive Manufacturing Innovation Institute. <http://namii.org/>

[62] EWI. Additive Manufacturing Consortium. < http://ewi.org/additive-manufacturing-consortium/>

[63] Lindemann C., U. Jahnke, M. Moi, and R. Koch. "Analyzing Product Lifecycle Costs for a Better Understanding of Cost Drivers in Additive Manufacturing." Proceedings of the 2012 Solid Freeform Fabrication

Symposium. <http://utwired.engr.utexas.edu/lff/symposium /proceedings Archive/pubs/Manuscripts/2012/2012-12- Lindemann.pdf>

[64] National Academy of Engineering. “Frontiers of Engineering 2011: Reports on Leading-Edge Engineering from the 2011 Symposium.” In National Academy of Engineering’s 2011 U.S. Frontiers of Engineering Symposium. Mountain View, CA. 2012

[65] Scott, Justin, Nayanee Gupta, Christopher Weber, Sherrica Newsome, Terry Wohlers, and Tim Caffrey. “Additive Manufacturing: Status and Opportunities”, March 2012. <https://www.ida.org/stpi/occasionalpapers /papers/AM3D_33012_Final.pdf>

[66] Hopkinson, Neil, “Production Economics of Rapid Manufacture.” In Hopkinson, Neil, Richard Hague, and Philip Dickens. *Rapid Manufacturing*. (Hoboken, NJ: John Wiley & Sons, 2006).

INDEX

A

B

C

D

E

F

G

M

N

O

P

Q

R

S

T